SUPERSTORM
1950

SUPERSTORM
1950

The Greatest Simultaneous Blizzard,
Ice Storm, Windstorm, and Cold
Outbreak of the Twentieth Century

David A. Call

Purdue University Press, West Lafayette, Indiana

Copyright 2023 by Purdue University Press. All rights reserved.
Printed in the United States of America.

Cataloging-in-Publication Data is available from the Library of Congress.
978-1-61249-796-9 (hardcover)
978-1-61249-797-6 (paperback)
978-1-61249-798-3 (epub)
978-1-61249-799-0 (epdf)

Cover photo courtesy of Weirton Area Museum and Cultural Center.

CONTENTS

AN INTRODUCTION

ON WEDNESDAY, NOVEMBER 22, 1950, CORDELIA WALKO OF ELYRIA, OHIO, looked at her enlarged abdomen and said a prayer. Her third pregnancy was nearly over, and she was ready. Due December 1, the baby growing inside her was causing increasing discomfort. An early delivery might improve her mood and make it easier for her to chase around her other children. Besides, the Christmas shopping season was just around the corner, and walking had become difficult.

A thousand miles to the east, a group of navy sailors prepared for a routine towing mission to transport destroyer escort 532 to Maine for recommissioning. The rough seas from a recent storm had finally settled down, and the weather forecast for the upcoming weekend called for quiet but cold weather. The task should not be difficult.

Finally, in Columbus, Ohio, Coach Wes Fesler of The Ohio State University worked on his game plan for the upcoming annual rivalry game against the University of Michigan. While his Buckeyes were highly favored against the unranked Wolverines, unrest was brewing due to the previous week's loss to a red-hot Illinois team and subsequent loss of the school's No. 1 ranking in the AP poll. Nonetheless, a defeat of their hated rival could secure both a trip to the Rose Bowl and a top ten ranking, and it would quiet the critics. Although Coach Fesler had failed to defeat Michigan in three prior tries, this time was expected to be different: his team was the favorite. A good game plan was critical to meet the high expectations.

What Coach Fesler, the sailors, and Cordelia Walko did not know was that their lives were about to be altered dramatically in the coming days by unforeseen and exceptional circumstances. Cordelia's baby would be delivered successfully, but in a snowdrift with a farmer's assistance. Waves of 50 feet would snap the sailors' tow lines and leave them adrift in the dark on a ship with no working systems. (Luckily, a courageous tugboat captain, Lt. William J. Bryan, rescued them amid the relentless storm.) And Coach Fesler's carefully constructed game plan was to be ruined by an unexpected blizzard, causing Ohio State to lose to Michigan and costing him his job. Ohio State would later gamble on a young, largely unknown coach from the western part of the state as his successor. That coach, Woody Hayes, built Ohio State football into a national powerhouse.

THESE EXTRAORDINARY EVENTS WERE ALL CAUSED BY ONE OF THE LARGEST, most intense storms to affect the eastern United States—Superstorm 1950. Commonly referred to as the Great Appalachian Storm or Great Thanksgiving Storm of 1950, this record-setter disrupted the lives of more than one hundred million people. Blizzard conditions and several feet of snow paralyzed Ohio, West Virginia, and western Pennsylvania, crippling commerce and shutting down major cities like Cleveland and Pittsburgh for the better part of a week. Altoona, Pa., experienced a complete blackout as record ice accumulations severed all high-voltage power transmission lines that fed the city. The Susquehanna River rampaged through Lock Haven, Pa., flooding more than three-quarters of the small city with more than 3 feet of water. Farther east, hurricane-force winds battered New England, New York, and New Jersey, causing coastal flooding and inland damage comparable to the Great New England Hurricane of 1938. The arctic air that accompanied the storm set new monthly low temperature records from Wisconsin to Georgia (many of which still stand to this day), wiped out late-season crops, and killed dozens from fatal fires. No state east of the Mississippi River was spared the wrath of the storm; the twenty-six states affected were home to approximately two-thirds of the U.S. total population. Damage was in the hundreds of millions of 1950 dollars (equivalent to at least $1 billion today), and 353 people perished.

At the time, Superstorm 1950 was the most expensive weather disaster in U.S. history, and its death toll the seventh highest for a nontropical storm in the United States. In the more than seventy years since the storm, the death toll has been exceeded only twice, both times by hurricanes. Superstorm 1950 simultaneously set records for snow, ice, rainfall, pressure, wind speed, and cold, and many of these records stand today. With its many hazardous facets, gigantic geographic area affected, and immense destruction and loss of life, this storm is best described as a *superstorm*.

But what exactly is a superstorm? Typically used to describe any large storm with multiple hazards, the term *superstorm* has no official definition. Misuse and overuse of the term are common, an issue that also plagues other such terms with similar meaning, such as *perfect storm*. This book works to change that by using Superstorm 1950 as a case study in what a superstorm is: a large cyclone with multiple hazards and major societal impacts.

The societal impacts of Superstorm 1950 were exacerbated by the lack of warning. Modern weather forecasts are accurate because of indispensable tools such as computer model simulations, weather satellites, and radar. A similar storm today would likely be predicted days or perhaps a week in advance. In contrast, computer models in 1950 were crude and used to study past weather events, not to look forward; weather radar was still being field-tested; and weather satellites (or any artificial satellites, for that matter) did not exist. Forecasters could use past analogues to forecast, but this storm did not follow the typical patterns of other storms. With no time to prepare, Superstorm 1950 surprised those affected. Tens of thousands of travelers were unable to return home, the sudden freeze damaged millions of dollars of mechanical equipment, and hundreds fled to higher ground when rising floodwaters threatened their homes and businesses.

The governmental response was very different than it would be to-day. In 1950, there was minimal federal involvement in disaster re-sponse. While states could petition Congress for relief, the process was slow and cumbersome. The Federal Disaster Relief Act of 1950, which gave the president the power to declare disasters, had been enacted prior to this storm, but it was too new to affect the response. Thus, the governmental response to Superstorm 1950 was almost exclusively ef-fected at the state and local levels. (In subsequent decades, the FDRA transformed disaster response from a state and local responsibility to a federal duty, culminating with the creation of the Federal Emergency Management Agency, or FEMA, in 1979.[1])

Superstorm 1950 inspired meteorological researchers of the time to find new methods and tools to improve forecasts, changing the practice of meteorology forever. At the time, Norman Phillips, Jule Charney, and other members of the Meteorology Group at the Institute for Advanced Studies at Princeton were rapidly advancing the science of meteorol-ogy by deriving the equations that describe the atmosphere's dynam-ics. The group was also experimenting with computer model simula-tions of the atmosphere using Department of Defense computers. The extreme weather and poor forecasting associated with Superstorm 1950 inspired these and future researchers to use the storm as a test case to refine their models. Each new model iteration was tested with this storm. Thus, Superstorm 1950 was critical in the development of the forecast models we use today—models that have revolutionized and greatly improved the quality of weather forecasts. It is not an exagger-ation to say that these improvements in weather forecasting have saved millions of lives across the globe.

BEFORE LEARNING MORE ABOUT THE STORM'S IMPACTS, WE WILL TAKE A trip back in time to 1950. Chapter 1 will provide an overview of what life was like in 1950 and highlight some differences between then and now. These differences will play out in the effects of Superstorm 1950, which varied based on race, gender, and class. Then in chapter 2 we'll learn more about the life cycle of Superstorm 1950, including how it formed, what made it so intense, why it followed such an atypical storm track, and whether it was predictable.

The bulk of the book, chapters 3–8, will describe the impacts of the storm. These are broadly grouped by disaster: heavy snow, ice, flooding, wind, and bitter cold. In chapter 9 we'll travel forward from 1950 to the present day by tracing the storm's long-term effects on meteorology and its critical role in developing today's forecast models.

Finally, in chapter 10, we will take a closer look at what exactly separates superstorms from ordinary storms by examining the effects of some other superstorms. We'll also look forward to see how the societal impacts from a comparable storm today would be similar in some cases (ice and coastal effects, for example) and quite different in others (such as snow and flooding). Understanding superstorms and their impacts will help society better prepare for future storms, ultimately saving lives, reducing property destruction, and lessening disruption.

But first, let's put on our leather jackets and vintage dresses and take a trip back to 1950.

PART 1

THE GENESIS

1

THEN (1950)

THE PRACTICE OF METEOROLOGY IN 1950 WAS TREMENDOUSLY DIFFERENT from that of today. There was less knowledge of the science, and modern tools such as satellites, radar, and computer model simulations were not available. These issues, combined with the unique nature of the storm, caused it to be poorly forecast.

But good forecasts are only one ingredient needed to reduce deaths and property losses from disasters. Governments and citizens need to take actions to protect themselves. Yet factors such as gender, race, and social class greatly influence one's ability to protect oneself. The Hurricane Katrina disaster in New Orleans vividly illustrates how social factors like race and age affect the vulnerability of individuals to disaster. Most victims were African American, and the mortality rates for African Americans were more than double those for whites. Almost half of all storm victims were age 75 and older.[1]

More than seventy years have passed since the November 1950 superstorm. In the same manner that forecasting now is different today than it was then, demographics and other aspects of daily life have changed.

POLITICS AND LIFE

In 1950, the thirty-third president of the United States, Harry S. Truman, was in office. The Cold War, a period of high tension between the U.S. and Soviet Union, was well underway, and fear of communists was pervasive in the United States. In January, Alger Hiss was convicted of two counts of perjury in connection with espionage for the Soviet Union. The following month, Senator Joseph McCarthy alleged that more than two hundred employees of the State Department were members of the Communist Party and part of a spy ring. In November and December, hundreds of newspapers, including those in major cities like Boston, Philadelphia, Miami, Minneapolis, and Los Angeles, ran a ten-part syndicated series on the "Threat of Red Sabotage." This series detailed how American Communists were actively working on a Soviet-directed campaign of sabotage and violence.

Another political item of note was the Korean War, which began on June 25, 1950, when North Korea invaded South Korea. The North Koreans captured almost the entire peninsula before American, South Korean, and United Nations forces rallied and drove them northward. By late fall, the allied forces had taken control of nearly the entire peninsula, and there was talk of bringing the troops home by Christmas. However, when thousands of Chinese troops attacked on November 25, coincidentally during the height of Superstorm 1950, the longest retreat in U.S. military history began. While those affected by the storm here were cleaning up, newspapers were filled with discussions about how to deal with China's intervention and if the atomic bomb should be used. In regions buried under heavy snow from Superstorm 1950, comparisons were made between the local snow and the harsh winter conditions the soldiers in Korea were simultaneously dealing with.

Domestically, the postwar economic boom continued. Builders were rapidly developing mass-produced suburbs, such as the various Levittowns found in multiple states. Income was on the rise, and retail stores were actively marketing television sets as Christmas gifts. Nonetheless, purchasing a TV was out of reach for most Americans.

Even the simplest 13-inch set cost around $200 in 1950 dollars (equivalent to approximately $2,000 today). Premium sets, as large as 16 inches and complete with record players and extravagances, were true luxury goods with prices close to $500 (around $5,000 in modern dollars). There wasn't much to watch on television either. Because of an FCC moratorium on new licenses, most larger cities had only one or two broadcast stations. Many smaller cities—and even some not-so-small cities such as Denver, Montgomery, Knoxville, Raleigh, and Scranton—had no TV stations. In fact, there were no stations at all in Colorado, Kansas, Mississippi, and South Carolina. Thus, the primary forms of news transmission in 1950 were the radio and the newspaper. In the aftermath of the November storm, many people stuck at home by heavy snow could receive news via radio only, as newspapers had great difficulty with both publishing and delivery.

Several innovations taken for granted today were rare or uncommon in 1950, and these influenced the impacts of the storm. Personal snowblowers had not been invented. Thus, dozens of (mostly) men died of a heart attack from shoveling snow. Credit cards were in their infancy; the first card, Diner's Club, had just been launched earlier in the year. A lack of credit caused cash shortages in Pittsburgh and other cities where heavy snow prevented banks from opening. Most homeowner's policies protected only against fire (unless riders were added for other perils), so homeowners had to pay for storm-related repairs out of pocket. Finally, most stores were closed on Sundays. This last difference was actually a good thing with respect to the storm, as it reduced traffic and made it easier for storm cleanup to begin.

GENDER, RACE, CLASS, AND DISASTER

Numerous factors affect a person's vulnerability to disaster. Some are physical, such as living near a chemical plant, while others are social, such as living in poverty. In the case of Superstorm 1950, the social vulnerability factors of gender, race, and class played significant roles in

---

CULTURAL HIGHLIGHTS OF 1950

The comic strip *Peanuts* and the animated movie *Cinderella* both debuted in 1950. So did the actor James Dean, whose first appearance was in a Pepsi commercial. In sports, the New York Yankees won their second consecutive World Series. They also won the next three.

---

which groups of people died or suffered great consequence as a result of the storm.[2]

Gender roles and expectations in 1950 were very different than they are today. Then, most working-age women were homemakers. According to the U.S. Department of Labor, in 1950 only 34 percent of women participated in the workforce, compared to more than 55 percent in the 2010s. The media encouraged women to stay home, to support their working husband, and to focus on raising children and maintaining a proper household. This could be seen in remarks by advice columnists such as Dorothy Dix, in advertisements, and in news stories and newspaper sections pitched directly to women. For example, in late 1950 numerous newspapers ran a fourteen-part syndicated series titled "How to Get a Husband," written by Cora Carlyle and based on her eponymous ninety-six-page book. (Apparently Ms. Carlyle's techniques worked well for her, as contemporary news accounts noted that she was "happily married" with two young children.) Perhaps the most atrocious example of sexist attitudes of the time was captured in this unsigned editorial from a newspaper in Butler, Pa., which may or may not have described a local event:

> Supposedly the marriage license clerk who suggested the bride and groom split the license fee was interested in seeing that both sides got off to an even start. There's merit here, but the man in the transaction might get awkward ideas to the effect that he is to be his wife's equal.
>
> —*BUTLER (PA.) EAGLE*, NOVEMBER 25, 1950

The blatant sexism seen in 1950 media is also evident in news reports describing the response to the storm. In Pittsburgh, for example, one news account reported that women were shoveling sidewalks around the YWCA yet still called them the "weaker sex."

Men were also constrained by gender roles. Shortly before the storm, a man near Scranton, Pa., accidentally snuffed out the pilot light on the stove while attempting to wash clothes, causing the near asphyxiation of his family from carbon monoxide. The *Scranton Tribune* blamed him for "playing the housewife role." Rigidity in gender roles caused many more men than women to die from Superstorm 1950. This was most common in snowy areas, where many men died from shoveling snow or traipsing through miles of snowdrifts. However, it also was seen in windy areas, where men were more likely to be shocked or injured while repairing buildings, and in cold regions, where men died in greater numbers from exposure to the cold and structure fires. In the South, race and poverty also played a major role in vulnerability.

Racial relations in 1950 were much worse than they are today, and the phrase "much worse" is likely an understatement. African Americans in the South were treated as second-class citizens. Millions left that region for better economic and social opportunities elsewhere in the Great Migration from 1910 to 1970. Those that remained in the South were segregated in all aspects of public life and were constantly at risk of harassment or worse for almost any reason. In the rare cases where charges were filed, all-white juries were quick to acquit perpetrators of even the most horrific crimes. When the federal government intervened to force a trial for a brutal police assault and blinding of Isaac Woodard Jr., a decorated African American veteran, an all-white jury in South Carolina took fewer than 30 minutes to acquit the local sheriff.[3]

At the time, segregation was strongly supported by many politicians, especially in the South. On November 28, 1950, the president of the University of Virginia (also a former governor) floated the idea of admitting African Americans to graduate and professional schools. The governor of Georgia rose in immediate opposition, and most other Southern governors simply remained silent. No one dared to support

the proposal. In Montgomery, Ala., where a new Teche Greyhound terminal was proposed, the architects provided twice as much seating in
the white waiting room (forty seats) as in the African American waiting room (twenty seats). Likewise, the restaurant for whites could seat
thirty-five, while the restaurant for African Americans had a capacity
of only fifteen. (The bus station was completed in 1951 and retired from
service in 1995. Today, it is the Freedom Rides Museum.)

Segregation was not just a problem in the Deep South. When the
University of Maryland floated the idea of a ballot referendum to allow
African Americans to be admitted, the *Baltimore Sun* used its Sunday
edition, with the largest circulation of any day of the week, to editorialize
against it. The editors argued that grade schools for African Americans
were now of equal quality to those of whites, and African American colleges were improving so quickly that "the goal of genuinely equal facilities under segregation . . . is within view." The landmark *Brown v. Board
of Education* U.S. Supreme Court ruling, which desegregated schools,
would not happen for another four years, and bus segregation was not
ruled unconstitutional until 1956.

Poverty rates for all U.S. citizens were much higher in 1950 than today. However, the amount of poverty among African Americans in
the South was truly astounding. In 1959, when data on poverty by race
were first collected by the U.S. Census Bureau, anywhere from 60 to
85 percent of Southern Blacks lived in poverty (rates varied by state).
Given the economic growth experienced in the U.S. during the 1950s,
the percentage of impoverished African Americans in 1950 was likely
even greater than in 1959.

Poverty and being non-white are both factors that increase one's vulnerability to disaster. With African Americans in the South having both
of these characteristics, it is no surprise that they constituted most of
the storm-related deaths in that region.

ULTIMATELY, DIFFERENCES BETWEEN THEN AND NOW IN THE SCIENCE OF
meteorology, as well as in society, increased the impacts of Superstorm
1950. The poor forecast gave government officials and the general

public little time to prepare. The next chapter details the storm's entire life cycle, describe why it became so intense, and explain why it was badly forecast. While every weather system is unique, Superstorm 1950 stands apart due to its intensity, atypical storm track, and record-setting weather.

2

THE STORM

"The result was the greatest wind force ever known over the region, the deepest snow ever experienced, the severest cold for the season, and mighty rainfalls that raised devastating floods."

—DAVID LUDLUM, WEATHER HISTORIAN[1]

SUPERSTORM 1950 WAS ONE OF THE STRONGEST MIDLATITUDE CYCLONES to affect the eastern United States during the twentieth century. Its lowest air pressure reading, 978 millibars (28.88 in.), is among the lowest observed for a snowstorm in that part of the country.[2] (Lower pressure readings are associated with stronger storms.) Incredible disruption occurred: paralyzing snow in Ohio, West Virginia, and western Pennsylvania; record cold throughout the Southeast; and wind damage throughout the Northeast.

Every year, many midlatitude cyclones form over or near North America, and most are unremarkable. In the case of Superstorm 1950, all of the necessary ingredients for an extreme storm came together perfectly to create a storm that winter weather experts Paul Kocin and Louis Uccellini would later call "the benchmark against which all other major storms of the twentieth century could be compared."[3]

NOVEMBER 1–23, 1950
THE ANTECEDENT CONDITIONS

Fall weather in the continental United States is highly variable. Lingering warmth from summer provides pleasant mild periods, but as the season progresses, cold spells from the north become more intense and widespread. As the calendar flips to November, the increasing contrasts between cold and warm create stronger, more intense storms.

While November is often variable and stormy, the weather in November 1950 was extreme even by November standards. The month began with widespread, record-setting warmth across the eastern United States. All-time monthly high temperature records were set in more than a dozen states, including places as far apart as Minnesota, Georgia, and New England. A few weeks later, a large number of these places set new all-time monthly record lows. In other words, many places in the eastern U.S. observed both their warmest *and* coldest temperatures ever for the month of November in the same year. Such extremes helped to fuel Superstorm 1950.

After the record warmth at the start of the month, the weather for the next couple of weeks was relatively typical for November—meaning highly variable—with cold spells, mild periods, and occasional storms bringing rain, and, in northern states, snow. Toward the end of the month, an unusually strong cold air mass (for November) developed over western Canada and started moving southeastward. On November 22, 1950, the cold air spilled into the northern Plains. Its presence was quickly felt: in Minneapolis, Minn., the temperature tumbled by forty-five degrees in 19 hours. The following day, residents of Minnesota, Wisconsin, and other nearby states shivered through one of the coldest Thanksgivings on record, with subzero cold common that night.

Unimpeded by physical barriers such as mountains, the record cold air mass continued to advance southward. Widespread temperatures below 0°F set all-time November record lows in Indiana and Ohio.[4] In

Cincinnati, the coldest temperature for all of 1950—even colder than those experienced the previous January and February—occurred on November 25, when the mercury fell to 5°F.[5] Further south, a balmy Thanksgiving was followed by sudden sharp decreases in temperature on the order of fifty degrees in 24 hours—and rare November snow in the Deep South. New record lows for the month were set in cities from Mississippi to South Carolina. Nashville, Tenn., had its earliest subzero temperature reading of the season and received a November-record 7.2 inches of snow. In Atlanta, the previous November record of 11°F was obliterated when the temperature fell to 3°F. Widespread natural gas shortages there caused pipes to freeze and factories to close. Even southern Florida was not spared, with new monthly records set in Fort Myers (34°F) and West Palm Beach (36°F). The new records shattered the old ones by an average of six degrees.

The map in figure 1 depicts the surface weather conditions on Thanksgiving evening, November 23, at 8:30 p.m. EST. Where is Superstorm 1950 on this map? It is not on there, because it has not yet formed. Weather aficionados know that storms are areas of low pressure (identified as "L" on a weather map), but the storm located over Lake Superior is not Superstorm 1950 and will weaken and disappear shortly. The

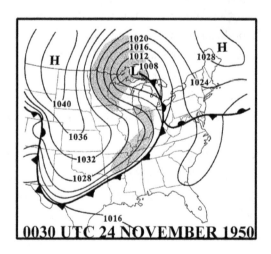

FIGURE 1. Surface weather map from 7:30 p.m. EST Thursday, November 23, 1950. (Map from Kocin and Uccellini [2004, 348], figure 9-27. © American Meteorological Society. Used with permission.)

0030 UTC 24 NOVEMBER 1950

leading edge of the cold air is depicted by the solid line with the triangles, or what meteorologists call a cold front, but the front has not yet reached most of the South. The high pressure ("H" on the map) responsible for the record cold is in central Canada and about to enter eastern Montana. Another high pressure is just east of Maine. Typically, we think of high pressure centers as bringing fair weather conditions, not storms, and while the eastern high pressure is no exception, it will become one of the most important factors in Superstorm 1950's later development.

FRIDAY, NOVEMBER 24

THE STORM RAPIDLY DEVELOPS

Figure 2 depicts the weather conditions on Friday, November 24. The top map is from the morning (12 hours after the map in fig. 1); the bottom map from is 12 hours later (or 24 hours after fig. 1).

The first thing to notice is the progression of the cold front. Overnight it passed through most of Ohio, Kentucky, Tennessee, and Mississippi; by evening it cleared the entire Gulf Coast as well as Georgia and most of the Carolinas, though the pace is slowing as the front moves across Pennsylvania and New York. Bitterly cold air followed the front, with temperatures quickly falling below freezing as far south as Biloxi, Miss., and Mobile, Ala., which was highly unusual for November. Further north, in Steubenville, Ohio, just west of the northern panhandle of West Virginia, the temperature plummeted from 38°F at midnight to 18°F at sunrise to 14°F at noon.[6] Measurable snowfall also accumulated in a very large area from Buffalo, N.Y., as far southwest as central Mississippi. In the South the snow caused great travel difficulties and inconvenience, while in the Midwest it disrupted daily life but primarily was a nuisance.

The first hint of Superstorm 1950 is seen over far southwestern Virginia in the top map in figure 2. Newly formed, it is very weak with a

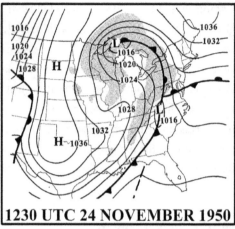

1230 UTC 24 NOVEMBER 1950

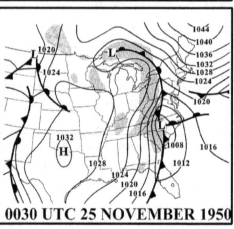

0030 UTC 25 NOVEMBER 1950

FIGURE 2. Surface weather maps from 7:30 a.m. (*top*) and 7:30 p.m. (*bottom*) EST Friday, November 24, 1950. (Map from Kocin and Uccellini [2004, 348], figure 9-27. © American Meteorological Society. Used with permission.)

pressure just under 1016 mb. Over the next 12 hours its pressure fell as it strengthened and moved into central North Carolina, as shown in the bottom map. Forecasters at the time were neither surprised nor concerned by its development. As strong cold fronts approach the East Coast, new low pressure centers often develop from the contrast between cold air over land and warm air over the Gulf Stream, a warm ocean current that parallels the East Coast. Typically, such a storm would then move out to sea—as this one was expected to do—or hug the coast and progress northeastward. This storm did neither, fooling the forecasters and catching the general public by surprise.

WEATHER MAPS TYPICALLY ENCOUNTERED ON TELEVISION OR THE INTERNET show conditions near the ground, or what meteorologists call "surface maps" since they show weather conditions nearest to the Earth's surface. We live on the ground, of course, so these conditions are most relevant to daily life, but conditions above the ground, or aloft, are often critical for determining whether storms will intensify or weaken. Conditions aloft are also helpful in assessing the steering winds that determine a storm's future path. Such insights into the state of the atmosphere are important for predicting future weather. Maps of conditions aloft make clear that this was no ordinary low pressure.

One of the most important maps of conditions above the ground is the 500-millibar chart, which shows weather conditions about 3 miles above the ground. Approximately half of the atmosphere, by weight, is below this level, and favorable conditions here are critical for the development of midlatitude cyclones, like Superstorm 1950. On a 500-millibar chart, solid lines are contours of height and show the altitude above the ground where the air pressure equals 500 mb. Since cold air is less dense than warm air (assuming equal pressure), smaller height values correspond to relatively cold air.

The 500-millibar chart from the evening of Friday, November 24, is shown in figure 3. An unusually strong upper-level low and exceptional pocket of cold air is shown across the Ohio Valley. Notice that the 500-mb heights over that region are the lowest of any location on the map, even lower than those in northern Canada.

Surface low pressure systems strengthen when 500-mb lows are upstream of them; stronger 500-mb lows result in faster strengthening. The magnitude and position of this 500-mb low relative to the surface low pressure, which was over Virginia at the time, are ideal for it to rapidly intensify.

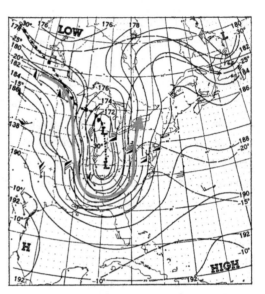

FIGURE 3. The 500-millibar chart as of 10:00 p.m. EST Friday, November 24, 1950. The gray arrow denotes the location of the jet stream. (Map from Smith [1950].)

Additionally, the polar jet stream, the river of fast-moving air aloft that separates warm from cold, was also favorably positioned. Caused by the contrasts between tropical and polar air, this narrow band of fast winds is typically around 35,000 feet above the ground; this is the height where jet airplanes fly in the atmosphere. The jet stream is often informally called the midlatitude jet stream because it is usually found between thirty and sixty degrees latitude. In summer, it is typically over Canada, while in winter it is typically found above the southern half of the United States, but such general statements obscure the considerable variations in location during most seasons.

The jet stream is thousands of feet higher than the weather depicted in figure 3, but I added its approximate location as a gray arrow. What's most notable is its position. It is highly unusual for the jet stream to be over southern Alabama and Georgia in November. Because of the strong contrast between the polar air over the Appalachians and lingering warmth from summer in the South, the jet stream was also unusually fast, with wind speeds above 150 mph.[7] These will both enhance the processes set in motion by the upper-level low and further intensify the storm.

A subtle but equally important feature on figure 3 is the northward bulge of the solid lines (or *isoheights*) over Maine and New Brunswick. Similar to the relationship between upper-level lows and surface lows, a strategically placed upper-level high pressure (or hint of one, as depicted here) can help intensify a surface high pressure center. Thus, the surface high pressure center over Labrador strengthened dramatically from 1030 mb (30.42 in.) to 1050 mb (31.01 in.) in the 24 hours prior to 10:00 p.m. EST on Friday, November 24 (fig. 3). Ironically, these favorable conditions aloft were caused in part by the unusually strong upper-level low pressure over Kentucky.[8]

Labrador is in the upper right corner of the map. Further southwest, the air pressure over northern Maine also dramatically increased during that time to anomalously high values for late November. The strengthening high pressure was critical to the storm's later track and ultimately responsible for the incredible amounts of precipitation that fell over Ohio, West Virginia, and Pennsylvania. While under normal circumstances the surface low pressure over eastern Virginia would either move eastward (out to sea) or northeastward (along the coast), this record-setting high pressure blocked both paths, directing the storm to the north, and later the northwest.[9]

SATURDAY, NOVEMBER 25

THE BOMB!

The conditions of the upper atmosphere were ideal for the nascent storm to rapidly intensify into one of the strongest November storms ever. In addition, the high pressure to its northeast blocked it from moving out to sea, causing incredible amounts of precipitation to fall over land. Forecasters at the time weren't able to predict this. With modern knowledge and more data, the forecast is obvious. The computer models of today, which are heavily used by forecasters, would easily simulate the storm's rapid development and move to the north. Such simulations also would predict that heavy precipitation, fueled by moisture from an atmospheric river, would fall across Pennsylvania, most of Ohio, and West Virginia.[10] Finally, modern computer models would predict a widespread windstorm over New York, New Jersey, and New England as a result of the growing difference in pressure, or pressure gradient, between the storm and the high pressure over eastern Canada. All of these events came to pass.

Figure 4 shows the positions of surface weather systems at 7:30 a.m. EST on November 25, 1950, or 12 hours after the bottom map in figure 2. The cold front over the Carolinas had pushed off the coast, but farther north it had stalled, and in northern Pennsylvania it was actually drifting westward as a warm front. This caused unusual conditions just ahead of it. In Erie, Pa., for example, temperatures rose from 20°F at 9:00 p.m. on November 24 to 41°F by 1:00 p.m. on November 25; at the time of the map, the temperature was around 33°F. The storm had moved almost directly north in the previous 12 hours to a location near Washington, D.C. More notably, its pressure had fallen to 992 mb

FIGURE 4. Surface weather map from 7:30 a.m. EST Saturday, November 25, 1950. (Map from Kocin and Uccellini [2004, 348], figure 9-27. © American Meteorological Society. Used with permission.)

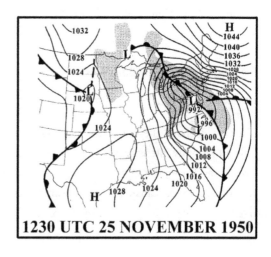

1230 UTC 25 NOVEMBER 1950

(29.29 in.), a decrease of 16 mb in the past 12 hours, for a total drop of 26 mb in the previous 24 hours. Modern meteorologists describe such explosive cyclogenesis as a "bomb."[11]

Meanwhile, to the northeast, the high pressure over eastern Canada was maintaining its strength as it drifted slightly southward. Caribou, Maine, in the northern part of the state, set a new November pressure record of 1042 mb (30.77 in.) just a few hours before the time of the map shown in figure 4.[12] The extreme difference in pressure between Caribou and Washington, D.C., is what caused the windstorm with hurricane-force winds on November 25 in the northeastern United States.

While the storm continued to intensify, the blocking high pressure over Canada led to the development of a unique frontal pattern in Pennsylvania, Ohio, and nearby areas, and highly unusual air movements. A map of the fronts from early Saturday afternoon is shown in figure 5. A new low pressure center had developed that morning near Erie, Pa., several hours before the time of the map. By 1:30 p.m. EST, as shown in figure 5, it had moved west-southwest to near Cleveland, Ohio. For a brief period the new low pressure coexisted with the old, but soon the new low over Ohio became dominant.[13] The pressure of the new low continued to rapidly drop as the storm became even stronger. Meanwhile, the stalled cold front began to advance again, but as cold air started to wrap around the low pressure from the south, the front now moved in a highly unusual direction. Because cold air originates near the poles and moves toward the equator, cold fronts in the northern hemisphere almost always move in a southward direction. In this case, however, the cold air was moving to the *northeast*. Perhaps stranger yet, on the other side of the new low a warm front was now advancing to the west-southwest across Ontario toward Michigan. The warm front eventually turned even more southward and entered Indiana from the northeast, an exceptional event. In other words, cold air around this system was moving from *south to north*, and warm air was moving from *north to south*.

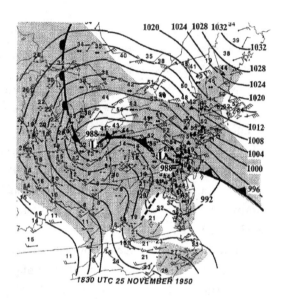

FIGURE 5. Surface weather map from 1:30 PM EST Saturday, November 25, 1950. (Map from Kocin and Uccellini [2004, 350], figure 9-28b. © American Meteorological Society. Used with permission.)

As the low near Cleveland started moving southwest and the "backward" fronts continued to advance, many bizarre weather phenomena happened. In Pittsburgh, Pa., temperatures fell from 21°F to 9°F after winds started coming from the *south*. Simultaneously, in Buffalo, N.Y., 200 miles north of Pittsburgh, the previous day's snow quickly melted under balmy fifty-four-degree weather despite gales from the *northeast* (which normally cause cold weather). In Detroit, heavy snow changed to rain as the temperature rose nearly thirty degrees during the day despite clouds and steady precipitation. The warmth in Buffalo and Detroit was short-lived. Late in the day, the cold front passed though (arriving from the *east* and moving to the *west*) and the snow and cold returned. In Buffalo, for example, temperatures that evening fell twenty-seven degrees in 6 hours despite winds now coming from the *south*, a direction almost always associated with warming conditions.

Meanwhile, residents of Indiana, who had simply experienced cold weather thus far, noticed storm clouds rolling in from Ohio as the storm neared. By evening, they too were experiencing blizzard conditions as 6 to 12 inches of snow blanketed the eastern two-thirds of the state and snowdrifts blocked roads statewide. By this point the air behind the warm front wasn't so warm anymore. Temperatures in Indiana failed to rise above 32°F the entire day, though this was considerably warmer than the widespread subzero readings from early morning.

While the fronts continued their atypical progress, the storm strengthened as it drifted further southwest into Ohio. The pressure continued to fall quickly (though not fast enough to still call it a bomb) and finally bottomed out at 978 mb (28.88 in.) over northern Ohio around 1:30 a.m. EST on November 26.[14] Dayton, Ohio, set a record for *lowest* November air pressure at 983.7 mb (29.06 in.), less than 24 hours after Caribou, Maine, set a monthly record for *highest* November air pressure.[15] Luckily, at the time the Dayton record was set the pressure in Caribou had decreased slightly, or the damaging winds in the Northeast might have been even more intense.

SUNDAY, NOVEMBER 26, AND BEYOND
THE SHORT, FINAL CHAPTER

By Sunday morning, November 26, the worst of the storm was over. The cold air was wrapped fully around the low, causing the cold front to catch up with the warm front and occlude. The southwestward-moving low had now moved directly under the upper-level low pressure center, an unfavorable location for further development. The high pressure formerly over Labrador had finally relaxed its grip; it was moving toward Greenland and weakening. Although light precipitation continued to fall from New England westward to Indiana on this day, the intense precipitation was over.

Over the next 24 hours the storm looped back to the east and then moved north across northeastern Ohio, crossing Lake Erie into southwestern Ontario. It then drifted westward and moved over Lake Huron, with its pressure quickly rising as it weakened. The connected occluded front slowly progressed eastward into the Maritime provinces.

Although the storm was no longer a threat, persistent cold weather and additional light snow (especially downwind of the Great Lakes) limited cleanup efforts early in the week. Additionally, leftover rainfall in New England would cause additional deadly flooding there. Quiet weather and more seasonable temperatures would not return until the end of the week.

THE TOLL

Superstorm 1950 left widespread destruction across the eastern United States. Incredible record-setting snow buried eastern Ohio, western Pennsylvania, and West Virginia; at the time of this publication, it remains the No. 1 ranked storm (by a large margin) in the Regional Snowfall Index for the Ohio Valley.[16] It ranks among the top ten snowstorms in the Northeast region despite only covering western Pennsylvania in heavy snow.

Figure 6 is a map of total snowfall; select amounts from within the hardest hit area are listed in table 1. Nearly every place within Ohio, West Virginia, eastern Kentucky, and western Pennsylvania experienced at least 12 inches of snow, and more than 24 inches fell in a 100-mile-wide band from central Lake Erie south through West Virginia. The Mountain State experienced the greatest average snowfall and highest individual totals. More than 40 inches of snowfall was reported in eleven counties, and in Harrison, Lewis, Marion, and Randolph Counties, some locations received more than 50 inches.[17] Only the Eastern Panhandle was spared.

Farther east in Pennsylvania, relentless rain caused flooding, especially along the West Branch of the Susquehanna River, and forty-eight river gauges reported water levels above flood stage. This might have been the biggest flood of 1950 for Pennsylvania, had not more severe flooding occurred the following week (though the later flooding was partially the result of saturated soil and swollen rivers left behind by

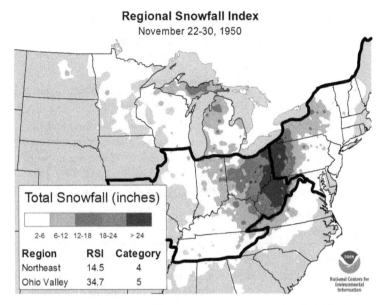

FIGURE 6. Map of total snowfall from Superstorm 1950. The thick black lines denote the boundaries of the Ohio Valley and Northeast regions. (Maps with numbers indicating accumulation amounts are in both Smith [1950] and Kocin and Uccellini [2004, 347].)

Superstorm 1950[18]). Precipitation totals from major Pennsylvania cities are shown in table 2. Most places in eastern Pennsylvania received more than 3 inches of rain, and Pottsville and Reading both received more than 6 inches. In west central Pennsylvania, in places like Clearfield and Altoona, much of this precipitation fell as freezing rain, where it destroyed the electric grid. In those cities, as well as in Bradford, snow is also included in the precipitation totals.

Finally, the windstorm in the Northeast was widespread and severe. Damage was compared with the Great Hurricane of 1938, and many places experienced wind gusts equivalent to those of a hurricane, or at least 74 mph. Select wind reports are listed in table 3. Hardest hit areas included eastern areas of Upstate New York and much of Vermont, where the storm was considered the greatest windstorm in recent history by utility officials.

TABLE 1. Total snowfall for November 25–28, 1950, for selected cities in West Virginia, Ohio, and Pennsylvania.*

Location	Snow (in.)	Additional context
Pickens, W.Va.[†]	57.1	Highest official storm total.[‡]
Parkersburg, W.Va.[†]	34.4	At the time, the snow from this event was greater than the most snow ever observed in an entire *month*.[§] Within one 24-hour period, an incredible 28.3 in. fell.[¶]
Clarksburg, W.Va.	34.0	An additional 4.0 in. fell on Nov. 29–30.
Charleston, W.Va.	25.1	
Bluefield, W.Va.	19.3	
Steubenville, Ohio	44.0	Largest snow total in Ohio from the storm.
Youngstown, Ohio[†]	28.7	17.0 in. on Nov. 25 is second biggest 1-day total in Youngstown history.
Cleveland, Ohio[†]	22.0	
Lima, Ohio	14.4	
Columbus, Ohio[†]	13.8	
Toledo, Ohio[†]	8.1	
Cincinnati, Ohio[†]	7.8	
Meadville, Pa.	35.3	Largest snow total in the state.
Pittsburgh, Pa.[†]	30.2	Greatest snowfall on record for Pittsburgh.
Erie, Pa.[†]	28.4	An additional 3.5 in. fell on Nov. 29–30.
Franklin, Pa.	26.5	Measurable snow fell 10 days straight, Nov. 21–30, 1950.

*Except where indicated by a dagger (†), totals were compiled from state climatological reports for the period Nov. 24–28 to be consistent with Smith (1950).

†Directly reported by Smith (1950).

‡A few modern sources list a 62-in. snow total in Coburn Creek, W.Va., but this value appears erroneous. This total did not appear in any accounts at the time or even those later in the 1900s, and it is not listed as the official W.Va. state record for snowfall. Snowfall totals and snow-to-liquid ratios at other nearby stations suggest that around 36 in. of snow fell at Coburn Creek.

§Jan. 1994 is the only month with more snow than Superstorm 1950.

¶Bristor (1951).

TABLE 2. Precipitation totals from major Pennsylvania cities for November 24–26, 1950.*

Location	Precipitation amount (in.)	Additional context
Bradford	3.23	Total includes snow.
Clearfield	4.33	Total includes freezing rain and snow.
Altoona	4.75	Total includes freezing rain and snow.
Williamsport	2.69	
Reading	6.09	Pottsville, 35 miles northwest, received 6.86 in., the greatest amount in the state.
Scranton	3.47	
Allentown	4.79	
Philadelphia	3.46	

*Totals are for rain unless noted otherwise. Cities are listed from west to east. Additional rain fell in many places after Nov. 26, but it was not significant.

TABLE 3. Select wind reports from the Northeast.*

Location	Fastest sustained wind (mph)	Fastest wind gust (mph)
Allentown, Pa.	n/a	88
Newark, N.J.	n/a	108
New York, N.Y.	61	94
Syracuse, N.Y.	50	90
Albany, N.Y.	66	83
Burlington, Vt.	63	75
Hartford, Conn.	70	100
Boston, Mass.	62	86
Concord, N.H.	n/a	110
Mt. Washington, N.H.	120	160[†]
Portland, Maine	58	76

*Wind gusts from state and national climatological reports and Ludlum (1982, 59).
[†]The anemometer was unable to report wind speeds above this value, so the actual winds were probably greater.

WHY THE STORM WAS SO EXTREME

To generate such an exceptional event, all meteorological factors had to come together perfectly. The first critical factor was the record cold air mass that penetrated far into the southern United States. Not only were previous November temperature records shattered, but the air aloft was also extremely cold. In the Carolinas, temperatures 1 mile above the ground were an astounding six standard deviations below normal, and those at 500 mb were four to five deviations below normal.[19] For context, any observation more than three deviations from normal is highly unusual. Since storms are driven by contrasts, having exceptional cold air throughout the atmosphere greatly enhanced the storm's intensity.

The second factor was the blocking high pressure center in eastern Canada. The peak pressure, 1049 mb (30.98 in.), was exceptionally strong, especially for November. As the high moved further east, the pressure over Greenland was more than three standard deviations above average for its latitude, while the central low pressure over Ohio was more than three standard deviations below average for its latitude. With such large departures from normal, contemporary researchers classified both of these as "very intense."[20] The intense high pressure blocked the storm from moving out to sea and actually pushed it northwestward across Pennsylvania and Ohio. Furthermore, it caused an extreme pressure gradient to develop, triggering hurricane-force winds over a large area. Ironically, part of the reason for such an intense high was the deep upper-level trough that fed the storm itself.

The third factor was an extremely strong jet stream. The exceptional cold air in the eastern United States helped intensify the jet stream to remarkable levels. Contemporary reports claim that it exceeded 200 mph.[21] (Generally any jet stream speed over 150 mph is very strong.) Unsurprisingly, winds at lower levels were three to five standard deviations above average, especially in New York and New England.[22] Strong winds at low levels also helped increase rainfall rates and amounts.

In the end, Superstorm 1950 was one of the strongest November storms to affect the eastern United States.[23] But it was not unprecedented. An analogue storm had occurred November 8–10, 1913, with a somewhat similar storm track. Nicknamed the White Hurricane because of the combination of extreme winds and heavy snow, this storm was the deadliest natural disaster to strike the Great Lakes. Dozens of ships were stranded or foundered, and more than 250 mariners died.[24] The White Hurricane also brought heavy snow to Cleveland (22.2 in.), western Pennsylvania, and West Virginia, where more than 36 inches buried Pickens.[25] However, little to no precipitation fell east of the Appalachians. The blocking high was considerably weaker as well, so the windstorm in the Northeast did not occur. Instead, the winds were caused by a strong high pressure centered over Minnesota, where pressure values of 1025 mb were 56 mb higher than those over the Great Lakes, helping to fuel extremely strong winds.[26]

Despite the similarity to the White Hurricane, forecasters in 1950 did not recognize Superstorm 1950 as an analogue storm. Simply put, they misforecasted it. This statement is not meant to condemn them; they simply lacked too many pieces of the puzzle to see the full picture. Not all equations that govern atmospheric motion were known or understood at that time, and real-time analysis of upper levels of the atmosphere was relatively new. Because weather satellites and computer model simulations did not exist, there was insufficient data over ocean and polar areas. With modern forecasting tools and methods, the 1950 superstorm could have been forecasted in advance.[27] Nonetheless, even if the storm had been correctly forecasted, it is unlikely that the forecasts would have been heeded. Weather forecasts in 1950 were given nowhere near the credence of those today. And even if the forecasts had been heeded, the magnitude of the weather was so extreme that government officials could not have done much to prepare.

That said, there is an upside to the poor forecast: It inspired meteorologists of the time to improve their understanding of the atmosphere and to develop computer model simulations. Today's accurate weather forecasts are the result of these model simulations.

WITH SUPERSTORM 1950, ALL CONDITIONS CAME TOGETHER PERFECTLY TO cause a record-setting, multi-hazardous, midlatitude cyclone over the eastern United States. Of course, what really makes this storm memorable is the great loss of life and damage it wrought on the country. The next six chapters describe these effects, with each one focusing on a different type of extreme weather. Chapter 3 examines how record-setting snows shut down Ohio, western Pennsylvania, and West Virginia. The slow recovery process is enough of a story itself to warrant an additional chapter—chapter 4. Moving eastward into west central Pennsylvania, chapter 5 looks at the record-breaking ice storm that plunged Altoona, Pa., into darkness for a week. In eastern Pennsylvania, as well as New Jersey and New England, flooding was a problem; this is the topic of chapter 6. These areas, but especially Vermont and nearby parts of New York, were also buffeted by hurricane-force winds; this is the focus of chapter 7. Finally, chapter 8 catalogs the impacts of the unseasonable record cold on the South. A full understanding of the record-setting varied impacts of the storm will help to clarify why it was no ordinary midlatitude cyclone, but a true *superstorm*.

PART 2

THE EFFECTS

3

FIFTY-SEVEN INCHES

*"Never in the history of Butler County has there been
such a paralysis of travel."*

— *BUTLER (PA.) EAGLE*, SATURDAY EVENING,
NOVEMBER 25, 1950

*"Cleveland was tighter than molasses. Nothing moved—
except feet."*

— *CLEVELAND PLAIN DEALER*, SUNDAY,
NOVEMBER 26, 1950

WHILE ALL ASPECTS OF SUPERSTORM 1950 WERE RECORD-SETTING, THE
snow was the most disruptive as it paralyzed a large, heavily populated region. Amounts unequaled before or since buried the upper
Ohio Valley, disrupting normal life for more than a week. The timing of
the storm, over the Thanksgiving holiday weekend, was notable since
many people were traveling—and thus became stuck in place. It was
also somewhat unexpected as November is not usually the snowiest
time of year. Pittsburgh, for example, received nearly ten times its normal November snowfall from this storm alone. The snow and its impacts were unequaled before and after.

THURSDAY, NOVEMBER 23

A COLD THANKSGIVING

Thanksgiving Day across the Midwest was colder than average. Temperatures had progressively cooled after mild weather the previous weekend, and highs that day were mostly in the lower forties, well below average. Skies were overcast, and a cold rain began after noon. As temperatures fell toward evening, the rain changed to snow from west to east. While forecasters did not expect the snow to be a hardship, they did warn the public that a period of unusual cold was anticipated for the weekend. In states such as Minnesota and Wisconsin, numerous record lows had been set on Thanksgiving morning.

As the bitter cold progressed from west to east, falling temperatures and the arrival of snow began to cause some trouble. The first storm-related deaths occurred near Chicago, where temperatures fell to a record 0°F just before midnight on Thanksgiving night. Robert G. Nilles, 26, of Mundelein, Ill., was killed when his car skidded and overturned. In Manteno, Ill., about 50 miles south of Chicago, Ethel McPherson, 83, died when her son-in-law struck an oncoming truck while attempting to pass another vehicle on a snow-covered four-lane highway. In northwestern Ohio, two chain-reaction crashes occurred in Toledo that evening as temperatures plummeted into the teens. One pileup involved eleven cars, and five automobiles were damaged in the other. Both crashes occurred on bridges, which froze before adjacent roadways.

FRIDAY, NOVEMBER 24

CONDITIONS SLOWLY WORSEN

Snow from the cold front quickly swept across the region overnight. Friday morning commutes were disrupted throughout the Midwest, causing many workers to be tardy in cities such as Louisville, Cleveland,

Youngstown, and Pittsburgh. Naturally, motorists had the greatest difficulty driving on hilly streets, which often contributed to crashes. However, the crashes were primarily fender-bender types. In exception, two
motorists in Louisville, Ky., skidded into the paths of oncoming trains
and were killed. These were the first deaths from the storm south of the
Ohio River. Snow arrived late in the day in West Virginia and western
Virginia, where problems in cities like Bluefield, W.Va., and Roanoke,
Va., became most acute during the late afternoon rush hour or evening.
In Wheeling, W.Va., Elliot Browning, 70, died of a heart attack while
putting chains on his car.

Further north, in the Great Lakes region, the intensifying storm disrupted air and boat travel. In Michigan, Leslie K. Straub, a 26-year-old
World War II veteran, and his passengers, two teenage brothers, were
killed when their plane crashed. Although the weather was reported as
good when they left Dowagiac, Mich., by the time they reached Benton
Harbor, 15 miles northwest, the snow was intense, causing them to lose
their bearings as they attempted to land. Intense snow also caused navy
ensign Richard Baumgartner of Michigan to crash his plane about 10
miles north of Meadville, Pa., or 25 miles south of Erie. The force of
the impact created a 6-foot-deep crater, delaying recovery of his remains until the snow melted and equipment was available to pump
out the water.

In Toledo, the tugboat *William A. Whitney* ventured onto Lake Erie
during the storm while pulling two mud scows. The high waves caused
the second scow to break free, and visibility became so low that the captain simply anchored 3 miles outside the Toledo harbor and waited for
help. The Coast Guard managed to rescue all seven men on board late
the following night. The scow that drifted away was lost, which was
surely a disappointment given its $200,000 cost (in 1950 dollars; equiv-
alent to $2 million today).

Road crews in western Pennsylvania and Ohio increased their efforts to fight the accumulating snow during the day but could not keep
up. In Pittsburgh, despite salting the Boulevard of the Allies, a major

artery, four times between 7:00 a.m. and 12 noon, crews could not keep it open. Conditions were no better 60 miles southwest, where road crews in Wheeling, W.Va., cleared U.S. Route 40 more than twenty times between 5:00 a.m. Friday and 4:00 p.m. Saturday before finally losing the battle with the snow.[1] As conditions worsened into Saturday, city leaders responded with all available resources, and Pittsburgh's and Cleveland's mayors vowed to continue working until the crisis was over. Neither mayor went home again until December.

Although the exceptionally cold air in this region was less deadly than in the southern United States, deaths still occurred, especially south of the Ohio River. Newspapers reported that seventeen people died from the storm over the weekend in Kentucky and West Virginia. Deaths resulted from carbon monoxide poisoning, fires from substandard or improper use of heating equipment, and exposure. There were also several close calls, where people were overcome by carbon monoxide and rescued in time. Although these deaths during the storm were tragic, many more deaths would occur after the snow ended and people attempted to dig out.

SATURDAY, NOVEMBER 25
THOUSANDS STRANDED

By Saturday morning, secondary roads had become impassible throughout the region. As the day wore on, road crews gradually lost the battle on all major arteries due to equipment breakdowns, unending snowfall, and general fatigue. The *Bluefield Telegraph* named three counties in West Virginia (Nicholas, Monroe, and Grafton) where all pieces of equipment failed, rendering road crews powerless. Making matters worse, falling and blowing snow buried the equipment in snowdrifts, where most of it remained until late the following week.

All vehicle traffic came to a halt as the snow continued. Vehicles slid into snowdrifts, broke down, or became blocked by other stopped traffic. With crime almost nonexistent, police focused on rescuing stranded motorists. In north central Ohio, the *Toledo Blade* reported that police rescued more than fifty people stuck on rural U.S. Route 6 west of Sandusky. Further south, near Williamstown, Ky., about halfway between Cincinnati and Lexington, U.S. Route 25 crosses through rugged terrain. Unable to reach stuck motorists by car, police commandeered a ten-car passenger train and slowly traveled down the adjacent rail line, rescuing more than three hundred stranded people, according to the *Louisville Courier Journal*.

Although it sounds like something from a comedy sketch, several out-of-town travelers became so disoriented that they actually confused streets with railway lines. Sadly, the consequences were no laughing matter. An Illinois man in Clarksburg, W.Va., was driving on railroad tracks when suddenly a train approached and his car stalled. Luckily a nearby fireman was able to flag down the oncoming train before tragedy occurred. The motorist escaped in time, but his car was damaged; the train itself was merely delayed. In Erie, Pa., Ernest Lytle, 65, was not so lucky. While crossing the railroad tracks on foot around 7:00 p.m. on Friday, November 24, in blinding snow, he apparently did not see the oncoming train. He was struck and killed instantly. Nearby witnesses were unable to tell police or the *Erie Times* exactly what caused him to be struck. Blizzard conditions with visibility near zero prevented them from seeing the crash.

With travel impossible, travelers, workers, and shoppers quickly filled downtown city hotels to capacity in large cities like Cleveland and Pittsburgh, as well as smaller ones like Butler and Oil City, Pa. Hotels had difficulty handling the crowds for two reasons: First, since weekends were normally less busy, the staffing levels were reduced. Second, impassible streets prevented some of their own workers from traveling.

People unable to secure lodging spent the night anywhere they could find a warm room and a flat surface to sleep on, including train station benches, church pews, bars, and even floors. In rural areas, stuck motorists crowded into residences, farmhouses, churches, and civic clubs. Roger Pickenpaugh's book on the storm's effects in Ohio, *Buckeye Blizzard*, contains dozens of stories on how communities and individuals did whatever necessary to help family, friends, neighbors, and strangers get through the storm.

Refugee camps of stranded travelers developed at smaller cities along major roads. Many of the stranded were football fans returning from rivalry games. In Kentucky, thousands of football fans returning from the Kentucky–Tennessee game in Knoxville, Tenn., became stuck along U.S. Route 25, with the greatest concentration in Richmond, Ky., about 25 miles south of Lexington. In Ohio, the Red Cross was dispatched to aid an estimated thirteen hundred University of Michigan fans and other travelers stuck in then-agrarian Union County, just northwest of Columbus. According to the *Columbus Dispatch*, around five hundred were in the county seat, Marysville, and another eight hundred were scattered along major roads elsewhere. Those stuck in town were housed in homes, churches, lodge halls, and restaurants, and fifteen voluntarily spent the night at the county jail. A few miles north of Marysville, thirty-one unexpected guests spent the night at Sam Westlake's farmhouse. Meanwhile, in Medina, Ohio, 30 miles southwest of Cleveland, more than five hundred truckers and other travelers unexpectedly spent the weekend. Police in Akron, 20 miles east, banned all vehicles without chains, causing 152 trucks to get stuck in Medina. Many additional travelers came from U.S. Route 42, which in 1950 was the main route from Cleveland to Cincinnati. A large group was housed at the YMCA, in addition to those in private homes and lodge halls. Travel on U.S. 42 did not begin moving again until Monday.[2]

The Pennsylvania Turnpike, the first long-distance limited-access highway in the country, was closed for the first time in its history. The road was closed from the western terminus in Irwin, about 20 miles east of Pittsburgh, to Bedford, about 85 miles to the east. This caused what the *Pittsburgh Press* termed the "greatest traffic snarl" in the ten-year history of the "all-weather" road. More than seven hundred motorists from thirty-two states and Canada were stuck in Irwin, unable to leave the turnpike as all roads leading away were snow-clogged. Another gigantic traffic snarl also formed 5 miles west of Irwin, in East McKeesport, the last populated town before Irwin. Here, an estimated one thousand travelers were stranded, unable to advance over hilly terrain to the turnpike entrance in Irwin (which was closed anyway) or to return home because of the fast-accumulating snow. Travelers in both towns spent the next few nights sleeping on church pews and theater seats, in fire halls and private clubs, and at private residences—anywhere they could find a place to spend the night. Meanwhile, local firemen scoured the area in search of food and other supplies. Although travelers were able to leave East McKeesport early the following week, the *Pittsburgh Press* reported that officials in Irwin were working to clear around 350 stuck or abandoned vehicles the following Wednesday, four days after the storm ended.

Travelers were not the only ones stuck in place. Boy Scouts on camping trips were also stranded by the deteriorating weather. In Pulaski County, Va., west of Roanoke, eight scouts were marooned on Friday night. On Saturday, they hiked to a nearby farmhouse, where half of the group spent the night while the remainder hiked to the scoutmaster's house to contact worried parents by phone. All of the scouts finally returned home, safely, on Sunday, according to the *Roanoke Times*. In Cleveland, fifty-four scouts were stuck in various metropolitan parks. They were rescued through a variety of means, including military truck, tank, and even helicopter.

The storm halted most forms of mass transit. On Friday, trolley and bus systems experienced delays but were able to maintain a semblance of normalcy. On Saturday, everything ground to a halt. In Cleveland, all sixty-four trolley routes stopped running. Pittsburgh's extensive trolley system, which in 1950 was the third largest in North America, behind only Chicago and Toronto, also came to a standstill on Saturday morning. This happened despite Pittsburgh Railways moving all sixty-six pieces of snow removal equipment to just six key routes. The collapse stranded more than 150 trolley cars sitting in the streets, further hindering traffic and snow removal operations. The transit system in less-hard-hit Columbus, Ohio, continued to operate for a bit longer, but operations still ceased by late Saturday.

Both intra- and intercity buses in all cities stopped running as well. Just like cars, they stalled or became stuck in drifts. Sometimes passengers attempted to push the buses out of drifts, with mixed results. In one ironic case, a group of passengers in Pittsburgh succeeded, only to see the bus continue on its way, leaving them stranded.[3]

Passenger trains, the primary method of moving people between cities, were the only form of transportation that continued to operate. Nonetheless, some were canceled, others were rerouted, and those that made it to their destinations were delayed by as many as 9 hours. In eastern Ohio, trains could travel only at the minimum allowed speed since poor visibility made it difficult to see the signals. This caused delays of up to 6 hours in this area, according to the *Columbus Dispatch*. Frozen and buried automatic switch boxes were another major problem, requiring trains to regularly stop so crew members could clear the switches by hand. Manpower shortages also plagued railroads as employees were trapped at home or stuck elsewhere. The New York Central Railroad avoided the worst of the weather by rerouting at least four Detroit to Buffalo trains through Canada, skipping all stations in northern Ohio (e.g., Cleveland), northwestern Pennsylvania, and southwestern New York State.

Poor visibility and an inability to keep long, drift-prone runways clear forced all airports in the region to close. Most ceased operations on Friday, earlier than ground-based mass transit systems. The Pittsburgh airport, for example, closed at 12:30 p.m. on Friday. Without modern deicing and plowing equipment, even airports on the fringes of the storm closed, such as in Roanoke, Va., and Lexington, Ky. Given the low number of airline passengers in 1950, the airport shutdowns had minimal effects.

As the storm raged on, municipal authorities everywhere, despite their best efforts, lost the battle to keep major arteries open. Desperate governors in Ohio, West Virginia, and Pennsylvania dispatched the National Guard. In Cleveland, tanks were used to provide aid beginning on Saturday, November 25. One 30-ton Sherman tank "cleared" snow by packing it down, allowing roughly fifty trapped motorists on the Eastern Shoreway to leave. Another came upon two stuck buses bound for suburban Berea. The passengers were consolidated into one bus, and the tank towed the bus 8 miles. The trip took 3 hours, according to multiple sources.[4]

For the next few days, tanks and other heavy National Guard equipment were the only motorized equipment moving through the streets of Cleveland and Pittsburgh. They continued to "clear" streets, perform rescues, and engage in other needed tasks. One of the funnier stories was detailed in the *Cleveland Plain Dealer*. A tank was dispatched to pick up an alleged dead body on a porch on Cedar Avenue in Cleveland. When the tank arrived, the noise awakened the corpse—which was actually a sleeping man. Opening his eyes to see a tank muzzle in his face, the man let out a scream and immediately fled the scene.

With motorized travel not available, a grocer in suburban Pittsburgh turned back the clock by using a horse to deliver milk and bread. A similar thing happened in Ohio, where a Chesterland farmer used horses to bring his milk to a Cleveland dairy.

As most people didn't own horses, foot travel became the only reliable option for getting around in snowbound areas. Many people walked extraordinary distances to get to work. Walking a dozen miles in fair weather is taxing for most people, but some residents did it in knee-deep snow. The president of the Cleveland Transit System walked 6 miles to get to work (a trip that took him 6.5 hours), while a fireman walked a distance of 12 miles from his house in Chesterfield to his station in Cleveland Heights. Roger Pickenpaugh chronicled multiple additional examples of Clevelanders walking great distances to get to work, to get to stores, or to check on relatives.[5] It is likely that this happened in Pittsburgh and elsewhere as well, even if it wasn't widely reported in the contemporary news media.

EVERYTHING IS CLOSED

HOW ROUTINE LIFE WAS DISRUPTED

The snowfall from Superstorm 1950 affected all aspects of routine life during the weekend of November 25–26 and following week. The remainder of this chapter will discuss how it affected life on Saturday and Sunday, which is when almost all snow fell in the snowbound region. The next chapter will discuss the disruptions after the snow ended. From business closures to sports to medical issues, no aspect of life was untouched.

BUSINESSES ARE CLOSED AND RELIGIOUS
SERVICES ARE (MOSTLY) CANCELED

Just like today, Christmas shopping and gift buying was a big deal in 1950. While Christmas creep has gradually extended the modern holiday season, in 1950 Christmas shopping still took place primarily between Thanksgiving and Christmas, with Saturday, November 25, being the first shopping Saturday. [6] Newspapers ran large advertisements before the weekend, and this would have been the biggest shopping day of the year had the storm not ruined the party. As the storm intensified and travel became impossible, major downtown stores in all cities had no choice but to the close early. There were simply not enough shoppers or employees to stay open. Thanks to the weather, almost all other businesses closed Saturday and remained closed on Sunday, if they were scheduled to open at all that day. The Cleveland Zoo managed to open on Saturday, but only four visitors showed up. The Pittsburgh Stock Exchange closed early on Saturday on account of no one showing up to trade. According to the *Charleston (W. Va.) Gazette*, this was its first unscheduled closing since a flood fourteen years earlier.

Most churches canceled Sunday services due to the weather. In exception, a few churches in urban areas managed to hold services if most parishioners were within walking distance. At some, attendance did not drop off much, but at most churches that opened, fewer than half of the usual number of congregants attended. In Youngstown, the *Vindicator* reported that area Catholic churches held mass, but the bishop relieved parishioners of their obligation to attend by virtue of "extraordinary" conditions. Only two protestant congregations in that city bothered to open. In Oberlin, Ohio, about 35 miles southwest of Cleveland, a humorous exchange occurred as a man attempted to clear sidewalks around First Church. As described by the *Cleveland Plain Dealer*, around 7:00 a.m., a (slowly) passing motorist shouted at the shoveler that "if the preacher were doing the shoveling, perhaps he wouldn't be so anxious to hold services," to which the man replied, "Brother, I AM the preacher!" The preacher was shoveling because the church's custodian was stuck in Cleveland.

Weddings, funerals, and other social events were also disrupted or canceled. Many Saturday weddings were simply postponed, while those that took place were delayed. In Pittsburgh, a bridegroom arrived at the church on time for his 10:00 a.m. nuptials, but the bride's vehicle, which had to travel only one block, got stuck on the way there. The bride walked home, changed into regular clothes, walked back to the church, and changed back into her gown, causing a 2-hour delay. In Clarksburg, W.Va., both the bride and groom made it to the ceremony, but only one usher arrived. The ceremony was still held, and the couple was so eager to proceed with the wedding that the bride forgot to remove her snow boots, according to the *Clarksburg Telegram*. By Saturday evening, conditions in Butler, Pa., had degraded so much that the bride and groom, their parents, and the minister used horses to get to the church for a 7:30 p.m. wedding, and still arrived 2 hours late. The caterer also showed up, but surely there were many leftovers. Only 20 of the expected 150 attendees—including most of the wedding party—made it to the reception. After the reception, the horses took the newlyweds and their immediate family back home. A bobsled was used to spirit the caterer to the nearest major road, and his car was the last one to travel down the road before it was shut down entirely. Finally, in Wheeling, W.Va., a Saturday wedding was postponed to Monday. Conditions on that day weren't much better, and most people missed the ceremony, including the best man. But the groom made it after walking through 16 miles of snow drifts. The walk took him 7 hours, noted the *Wheeling News-Register*.

Hundreds of funerals were also canceled as a result of the weather, and all interments were postponed. One poor soul in Wheeling, W.Va., nearly missed his own funeral, scheduled for the Monday after the storm. Mourners arrived to discover that the dead man's corpse was not there because of poor road conditions. The funeral was postponed.

NO NEWS IS NOT GOOD NEWS:
THE EFFECTS ON THE MEDIA

Newspapers, one of the most important sources of information in the pretelevision era, were of critical importance during the crisis. Unfortunately, they faced two major problems: difficulty in getting employees to work and difficulty distributing copies. Although people were not in the dark literally speaking (electric service was, by and large, maintained), without the news they were unaware of the larger state of affairs.

Where the snow was heaviest, newspapers had to cut back. In Pittsburgh, the *Pittsburgh Press* skipped Saturday evening's edition and focused all of its efforts on Sunday. (The *Press* even included written synopses of missing Saturday comics to give readers story continuity.) Although its main competitor, the *Pittsburgh Post-Gazette*, was not published on Sundays, weather conditions prevented it from publishing on Monday as well. In Cleveland, despite only having about one-sixth of the normal workforce on Saturday, the *Plain Dealer* managed to publish, although the editions were thinner than normal. Neither of its two competitors had sufficient staff to maintain continuity. Despite the lack of workers, the *Wheeling (W.Va.) News-Register* managed to print, but across town the *Wheeling Intelligencer* suspended operations for the first time in its 98-year history. Farther north, the *Youngstown (Ohio) Vindicator* skipped three days of newspapers (Saturday, Sunday, and

Monday), choosing instead to present "radio editions" on WMFJ. In an editorial, the publisher explained the decision:

> There seemed little use in making a herculean effort . . . merely to sustain a 75-year-old tradition [of daily publication]. . . . The real reason why *The Vindicator* did not publish is that it could not be delivered.

Most other larger newspapers in the region, including those in places like Columbus, Ohio, and Erie, Pa., managed to publish, though it should be noted that these places received less snow. (Many papers in small towns or rural areas only published on weekdays, which fortuitously allowed them to avoid the worst of the weather.) Although delivery was challenging for obvious reasons, nearly every newspaper in the region lauded the efforts of the carriers to deliver the paper, heaping praise on their youthful delivery workers for their tenacity.

CHATTERBOXES: THE EFFECT OF THE STORM ON UTILITY SERVICES
Although there was wind with the storm, it was not sufficient to disrupt electric and telephone service like it did in the ice- and windstorm-affected regions. Scattered outages happened in Cleveland, Ohio, and in Roanoke and Lynchburg, Va., but the main storm-related problem was getting crews to locations in a timely manner. Gas usage was also high, but with most industrial users closed owing to the holiday and the weather, shortages were not a widespread problem except in Kentucky. In that state, record usage caused some isolated problems with low pressure in Louisville, Lexington, and Somerset (about 70 miles south of Lexington). Problems in Somerset were severe enough that, according to multiple reports, some people were housed temporarily in city school buildings.

Elsewhere, gas problems resulted from special local circumstances. In Toledo, Ohio, there were many complaints of low gas pressure. A new 20-inch high-pressure gas line was supposed to have been finished in the summer, but material shortages delayed its completion.

City officials blamed residents for installing gas appliances without permits (a similar complaint was made by officials in Atlanta, Ga.), and they threatened increased penalties for illegal devices. In Grafton, W.Va., a bulldozer clearing snow ruptured an 8-inch gas main at approximately 7:00 p.m. on Wednesday, November 29, cutting off heat for approximately seventy-eight hundred residents. Although the repairs were completed by 11:00 p.m., the gas was not turned back on until 13 hours later, causing numerous hardships. The gas company blamed the delay on federal regulations that required employees to notify all fifteen hundred customers to shut off their appliances and noted that they had sent out fifty canvassers the following morning. According to the *Charleston (W.Va.) Gazette*, the mayor remained dissatisfied, especially since he had offered the use of his police force for assistance.

Isolated in their homes, or unable to reach their homes from elsewhere, people used their telephones in record numbers. New highs for phone use were established across the region. In western Pennsylvania, Pittsburgh operators were flooded with calls, while residents of Butler, Franklin, and Oil City were asked to limit their calls to no more than 3 minutes and to avoid nonemergency calls. In Ohio, call volume records were set in Cleveland, Tiffin, Fremont, and Columbus, while the Youngstown exchange had one of its biggest days in history. The story was the same throughout West Virginia, Kentucky, and western Virginia; newspapers in these states noted record usage.

Although some areas had automatic local dialing, in 1950 operators were needed for local calls elsewhere, and they were needed everywhere for long-distance calls. Because of high call volume, many operators worked double shifts, and extra operators were called in to work if available. With road conditions poor at best, companies sent out service trucks to bring the operators to their facilities. Most companies housed operators in downtown hotels. In Pittsburgh, the phone company rented more than 150 hotel rooms, while in Wheeling, more than a hundred operators were housed in two downtown hotels. While some customers had difficulty placing calls by reason of high call volume, service was generally maintained thanks to the operators and line crews. In rural West Virginia, maintaining the lines meant dispatching some linemen by foot to keep an eye on rural, isolated lines. The *Wheeling News-Register* also noted that residents were expected to provide food and housing to these workers.

"SHOOT TO KILL" ORDERS AND THE EFFECTS ON CRIME

The poor weather didn't just prevent employees, parishioners, and others from traveling. It also kept criminals at home, much to the relief of local police, who were overwhelmed by medical emergencies, stuck cars, and other problems. In Pittsburgh, local newspapers observed that no crime reports were made from 10:00 p.m. Friday, November 24, through noon on Saturday, November 25. In Youngstown, Ohio, police only made four arrests over the weekend, all for minor offenses. Two of the arrests were in connection with sleeping in building entrances, which could have been connected with the snow and cold. (The other two were for intoxication.) The crime blotter in Lexington, Ky., was similar. A sergeant quoted in the *Herald* described the weekend's crimes as "nothing but a few drunks anxious to get in out of the cold."

While crime in Columbus, Ohio, was also below average, an auto theft was reported on Saturday, but the storm ultimately exacted vengeance for the victim. The story began on Saturday afternoon when Robert Flory, a worker at a service station, was driving home from work. His car became stuck in the quickly deepening snow, and as he struggled to free it, two young men offered to help. One man drove the car while the second man helped Flory push it. As soon as the car was freely moving, however, the second man jumped inside and the two thieves drove away. Making matters worse, the day's sales receipts, totaling about $400 in cash and checks, were in the glove box. Flory was surely upset, but three days later he received great news. Police found the car, abandoned and stuck in a snow drift, not far from where it was originally stolen. Even better for Flory, the $400 in cash and checks was still in the glove box, untouched.

THE STRANGEST VEHICLE THEFT?

Another unusual vehicle theft, which may not have actually been a theft, occurred near Erie, Pa. A Cleveland-bound trucker hauling tens of thousands of dollars' worth of electric motors skidded off the road. He left the vehicle to get help. Meanwhile, police came along, dug out the truck, and handed the keys to the man assumed to be the driver, who immediately departed. The real driver returned soon after to discover his truck missing and promptly reported it as stolen. Six days later, the truck was discovered at a competing truck company's freight terminal in Cleveland. The FBI and everyone else involved were baffled as to how it got there and debated whether the man assumed to be the driver was a Good Samaritan or simply a thief.[7]

In contrast with every other city, a crime spree occurred in Cleveland. Although there were no homicides or shootings, dozens of smash-and-grab burglaries occurred on Saturday, leading to the arrests of more than a dozen alleged perpetrators—sources disagreed on the exact number. Some looting was perpetrated by opportunistic thieves stealing items such as televisions and jewelry, while others took more basic necessities such as bread, milk, meat, baked goods, and liquor. Thieves also stripped numerous cars. Overwhelmed police called in the National Guard to help patrol the main business districts. With the military now backing up the police, deputy inspector James E. McArthur promptly issued an infamous "shoot to kill" order for any policeman or guardsman who encountered a thief or other prowler; this harsh order made national news. The show of force, or, more likely, the deteriorating weather, put an end to the problems. The final tally of burglaries was similar to that of any other weekend.

Some prison sentences ended during the storm, but several prisoners refused to leave on account of the weather. In Cleveland, the *Plain*

Dealer reported that the county jail superintendent, short on labor, put freed prisoners to work in exchange for housing them for several days. In Fayetteville, W.Va., about 35 miles southwest of Charlestown but roughly 60 miles by car, three inmates scheduled for release on November 26 asked to have their stays extended rather than trudge for miles through snow that was several feet deep, according to the *Charleston (W.Va.) Gazette.*

THE "BLIZZARD BOWL" AND OTHER FOOTBALL DISRUPTIONS
In 1950, only thirteen professional football teams existed, and the All-America Football Conference had just folded during the previous year. College football was the most popular form of the sport, and rivalry games at the end of the season were among the best attended and most intense games of the season. High school football was also popular, and several championships or other notable games were scheduled for the days after Thanksgiving.

Officials canceled or postponed most high school games. In West Virginia, the Class B championship game pitting Poca and Vinson was postponed, and the Weirton–Stonewall Jackson tilt was canceled. In Cleveland, the *Plain Dealer*'s annual charity game between St. Ignatius and Benedictine high schools was postponed, the first such postponement in the twenty-year history of the game. Though the game was initially postponed to December 2, poor field conditions caused a second postponement to December 9.

In college football, Penn State and the University of Pittsburgh (Pitt) canceled their annual freshman football game less than 3 hours prior to kickoff on Friday, November 24, and then postponed their primary game, originally scheduled for Saturday, November 25, in Pittsburgh, to the following Monday. This was the first postponement in series history (dating back to 1893). The postponement was done so Penn State's football team, already in town, would not need to leave and make a second trip back. However, when faced with the prospect of clearing an estimated two thousand dump truck loads of snow from Oakland Field with no trucks available because they were needed elsewhere, organizers had no choice but to reschedule the game to Saturday, December 2. Stymied by a continued lack of trucks, later in the week the game location was changed to Forbes Field, which needed to be cleared anyway for the Pittsburgh Steelers game scheduled for Sunday, December 3. Ultimately, the Penn State–Pitt game was played on December 2 before seven thousand fans (as opposed to the thirty thousand anticipated the previous week). Almost all fans walked to the stadium as street parking was nearly impossible in the snow-strewn areas around the stadium. The lucky ones who attended witnessed a dramatic and close game. Pitt, down by 21 points, came back to make the score 21–20 but lost on a missed extra point try. It was the closest final in the rivalry since a 0–0 tie in 1921.

In Columbus, Ohio, the annual rivalry game between the University of Michigan and Ohio State went on as scheduled on Saturday, November 25, despite "weather conditions even an Eskimo would have called atrocious" according to the *Columbus Dispatch*. The absurdity of playing in such poor weather was obvious before the kickoff. The frigid 10°F temperature froze the tarp protecting the field so thoroughly that workers could only remove small strips at a time, and dozens of fans had to come down from the stands to help remove the final sections.[8] The tarp problems delayed kickoff by 20 minutes. In spite of the terrible weather, 50,503 out of 82,700 fans still showed up, though that number was probably a bit inflated.[9]

Once the game began, the abysmal playing conditions forced both coaches to abandon their game plans. Thanks to gusty winds whipping snow around the stadium and players' hands going numb, it was very difficult to complete any passes. Running the ball was no easier. Cleats provided no traction on the frozen surface, making it challenging to run and impossible to cut; in desperation, the Ohio State team switched to tennis shoes at halftime, with no better results. The cold took a toll on players as well. Ohio State halfback Vic Janowicz described his experience to the *Columbus Dispatch* after the game, saying, "It was like a nightmare. My hands were numb and blue. I had no feeling in them at all. I couldn't hang on to the ball."

All scoring in the first half of the game came from blocked punts. Ohio State blocked a Michigan punt and kicked a field goal soon after. Remarkably, these were the only points scored by either team's offense in the entire game. Later in the first quarter, Michigan's defense scored 2 points from a safety to make the score 3–2. Right before halftime, Michigan's defense fell on a blocked Ohio State punt in the end zone for a touchdown. This gave them a 9–3 lead.

The two universities' bands played at halftime. Both the cold and wind were problems—for example, Ohio State's formation included two large beach umbrellas—but the snow on the field was a bigger problem, as it obscured the lines. Nonetheless, the shows went on, and photos from the day show the formations were successful. Given the weather, Ohio State's formation of a dancing Hawaiian hula girl was very ironic.

Although it seemed as if the weather couldn't get any worse, it degraded further in the second half of play as the swirling wind-whipped snow intensified and the visibility dropped close to zero. The dwindling number of spectators in the stands gradually drifted into the lower stadium sections simply to see the action—when they weren't busy hurling snowballs at each other. With their offenses rendered impotent, the two teams began playing a game of hot potato by punting the ball away on second and even first downs. (In football, teams normally do not punt until fourth down.) Michigan's twenty-four punts accounted for more than one-third of its offensive plays.[10] Neither team's defense scored again, so at the end of the game, Michigan was the winner by the same 9–3 score as at halftime. Michigan won the game despite not completing a single forward pass or gaining a single first down, the latter of which has never occurred in any other official college football game.

Ohio State athletic director Dick Larkins took full responsibility for the game being played over both head coaches' objections. Allegedly he was goaded into the decision by Michigan's athletic director, Fritz

Crisler, who pointed out the difficulty in refunding tickets should the game be canceled.[11] Crisler may have also argued that had the game been canceled, Ohio State would have been named champion of the Big Ten Conference by default, which would have been a stain to its reputation.[12]

The disastrous third down play before halftime that gave Michigan the lead was widely panned by fans and the local media. Critics in the *Columbus Dispatch* remarked that the blocked unnecessary punt, as well as Ohio State's allegedly "lackluster" play, were evidence that Ohio State coach Wes Fesler should be fired. While Ohio State finished the 1950 season with its third consecutive winning record, Fesler had failed to defeat hated Michigan on four occasions, losing three times and tying on the fourth. The 1950 loss was particularly aggravating as it cost Ohio State both the conference title and a Rose Bowl bid. Fesler, who had contemplated resigning after the previous season, bowed to the pressure and resigned on December 8. He later coached at the University of Minnesota for three years before leaving coaching entirely. Fesler's resignation, a direct result of the disastrous loss in the "Blizzard Bowl," changed Ohio State football forever.

A HUNDRED MILES SOUTHWEST OF COLUMBUS, IN CINCINNATI, OHIO, AN-other college football game was played. This game, between the University of Miami (Ohio) and the University of Cincinnati, was played in what the *Cincinnati Enquirer* described as a "raging snowstorm" with a frigid 10°F temperature. Around ten thousand spectators watched a punt-filled contest rife with turnovers, though both teams did manage to complete forward passes and gain first downs. The final score was highly lopsided. Underdog Miami resoundingly crushed bowl-bound Cincinnati, 28–0, to win the Mid-American Conference. The upset victory earned Miami a bid to the Salad Bowl (yes, that was the actual name of the bowl) in Tempe, Ariz., and the victory also captured the attention of the larger university in Columbus. On February 18, 1951, Ohio State gambled on Miami's little-known, young (38-year-old) head coach by selecting him to lead its program. He was chosen over several more experienced applicants, and the fan base was wary of the new hire. In the end, the gamble was an overwhelming success: Miami's former coach, Woody Hayes, later led Ohio State to five national championships and turned the program into a national college football powerhouse. If Superstorm 1950 had not occurred, it is doubtful that the chain of events that brought Woody Hayes to Columbus would have occurred. In that alternate universe, Ohio State might still be a mediocre Big Ten team that regularly lost to Michigan.

As a postscript, despite losing its head coach, Miami of Ohio continued to be highly successful. Woody Hayes's successor, Art Parseghian, led the school to conference championships in 1954 and 1955 before eventually achieving national recognition at Notre Dame. Bo Schembechler's first head coaching gig was also at Miami of Ohio (1963–1968). Schembechler is perhaps best known for his time at Michigan, where he annually squared off against Woody Hayes.

THE SKY IS FALLING!

Coach Fesler's career was not the only thing to collapse as a result of the storm. Roofs collapsed as well. Any time a roof collapsed it was widely reported, but collapses at residences were rare: one in Erie, Pa., another in Anmoore, W.Va., and a third in Clarksburg, W.Va. More common were collapses of garage roofs, with at least ten of these. In addition, around a dozen roofs collapsed on large open structures such as factories, meeting halls, and stores. Some other minor collapses, like those of porch roofs in Pittsburgh, also occurred.

Some roof collapses occurred in Ohio and West Virginia, but they happened most in Pennsylvania, especially toward the west central part of the state, where rain mixed with snow. This caused a sticky wet snow that clung to surfaces and also retained rainwater. About 50 miles east of Pittsburgh, near Indiana, Pa., roofs collapsed at two roller rinks, one in Plumville and other in Blairsville. In Kane, Pa., about 90 miles north of Indiana, Pa., and also located near the rain/snow transition, the roof at the Truskan Products Company collapsed, causing $20,000 in damage. About 1 foot of heavy, wet snow mixed with extensive rain (around 3–4 inches), and the combined weight proved too much for the rafters to support. Further west in Erie, a 60- by 75-foot section of the roof over Rainbow Gardens, a large dance hall located just west of the city, fell in, causing an estimated $12,000 ($120,000 today) in damage. The collapse actually occurred on the Wednesday following the storm, and the *Erie Times* also noted that high winds were a contributing factor.

Fortunately, no human deaths resulted from any roof collapses, and only one injured person was reported: a 57-year-old man near Franklin, Pa.[13] However, more than two hundred turkeys were smothered when the roof of a turkey pen collapsed 15 miles east of Pittsburgh near Center, Pa. Unfortunately, roof collapses were not covered by homeowner's insurance, which was very limited at the time, though automobiles in garages were protected by comprehensive automobile insurance.

OH, BABY! (AND OTHER MEDICAL RESCUES)
The stork waits for no one, and a record-setting blizzard wasn't going to stop babies from being born. As travel conditions deteriorated on Saturday, police found themselves frantically chauffeuring laboring women to hospitals by all means. Throughout the snowbound region, more than a hundred women were transported in police cruisers, fire trucks, ambulances, National Guard vehicles, and even an air force helicopter.

One of the more exciting stories occurred in Columbus, Ohio, where a relay race with four vehicles attempted to transport Maggie Snelling, 24, to the hospital. The *Columbus Dispatch* detailed how initially a fire truck was dispatched to take her, but its water pump failed on the way to her house. A police cruiser was sent next. It successfully picked her up but became stuck on the way to the hospital. Another cruiser arrived at the scene and continued the journey, only to suffer from a mechanical failure when it was forced to detour around stopped traffic. Finally, a third police cruiser completed the race, getting the mother to the hospital 1 minute before a healthy baby boy was born. While Ms. Snelling (barely) made it to the hospital on time, police officers in Toledo, Ohio, and Erie, Pa., delivered babies themselves.

Also in Columbus, police carried a laboring woman a half mile when their vehicle became stuck in the snow. But overtaxed police forces could not help everyone attempting to reach a hospital. In Cleveland, Pittsburgh, Wheeling, W.Va., and elsewhere, newspapers outlined how husbands, neighbors, and random volunteers pulled sleds or carried women in makeshift slings and gurneys to area hospitals—sometimes laboring over distances greater than a mile. In Akron, Ohio, one woman somehow walked a half mile to the hospital on her own while in labor. She gave birth soon after arriving.

Other laboring women failed to make it to the hospital at all, and police were unable to reach them. Twelve miles outside of Wheeling, conditions were so bad that no rescue could be attempted. Instead, a local doctor spent hours on the phone helping with the delivery (which was successful). Another home birth occurred in rural northwestern Ohio, where twenty-two people were stranded at a farmhouse on the main road between Toledo and Fort Wayne, Ind. Although an ambulance in Hicksville, Ohio, was dispatched, it took the vehicle more than 2 hours to travel 5 miles, and it arrived 15 minutes too late. South of Elyria, Ohio, Frank and Cordelia Walko attempted to go to the hospital during the height of the storm (November 25), but got stuck in the snow a mile from home. A neighbor then drove them 4 miles closer to town before his car became stuck. The neighbor and Mr. Walko attempted to carry Mrs. Walko to the nearest farmhouse, but both repeatedly stumbled in the deep snow. She ended up giving birth to a healthy baby girl while in the snowdrift. Finally, in Franklin, Ind., one farm couple hosted *two* births over the weekend as two of their neighbors got stuck while attempting to reach the hospital.

Snowbound rural roads caused a home birth four days after the storm ended. In a rural area near Steubenville, Ohio, Mrs. Edward Taylor, a 28-year-old mother of two young children, gave birth in her snowbound farmhouse at 6:00 a.m. on Wednesday, November 28. Her husband had left two days prior to help clear roads, and he not been home due to fresh snowdrifts caused by breezy conditions. Arriving home at 9:00 that evening, he found his wife passed out from exhaustion and his new child on the floor. Fortunately, he was able to summon much needed medical attention, and the *Youngstown Vindicator* reported that mother and child were in excellent condition afterward.

Extraordinary rescues were required for other medical emergencies as well. Several cases of appendicitis necessitated immediate action. In Bourbon County, Ky., a 6-year-old boy was struck on Sunday. He was placed on a sled and pulled 2 miles to Lexington Pike, where an ambulance then took him to a hospital. Appendicitis also struck 13-year-old Shirley Sloan of Pittsburgh during the height of the storm. In a 4-hour ordeal, rescuers used a sled, two police cars, and an ambulance, then finally carried her on foot to the hospital. More than fifty policemen, firemen, and members of the local American Legion Post aided in the rescue effort. A 19-year-old woman in Pittsburgh was stricken with appendicitis also. Two sleds and a group of mostly strangers pushed, pulled, and carried her through knee-high drifts. In west central Pennsylvania, Mrs. Robert E. Nimmo was visiting her parents near the small city of DuBois, where a mix of snow and ice had fallen. Struck by appendicitis, she was carried a half mile from their snowbound house to a waiting ambulance. As doctors prepared to operate at the hospital an hour later, the power went out, forcing them to perform the operation by flashlight. Luckily it was successful, and the *Butler (Pa.) Eagle* described her prognosis as good.

Hospitals throughout the region were overwhelmed. Besides the usual births and emergencies, the patient load was increased by victims suffering from heart attacks, hypothermia, or injuries from falls. The poor weather also prevented hospitals from releasing healed patients, creating a shortage of available beds in cities like Cleveland and Pittsburgh. Staffing shortages caused by the weather further compounded the problems. Doctors and nurses quickly found themselves pressed into other forms of service. At Presbyterian Hospital in Pittsburgh, one surgeon found himself slicing beef as fellow doctors cooked, served meals, and washed dishes. Another surgeon manned the elevator. In Youngstown, Dr. Henry Shore hiked through knee-high snow to reach the hospital. After completing his rounds, he washed pots and pans while a chaplain peeled potatoes. To reduce patient load, Cleveland hospitals postponed all elective surgeries, creating a backlog for weeks. (This likely happened in other cities, too.)

SUPERSTORM 1950 DUMPED INCREDIBLE AMOUNTS OF SNOW ON WESTern Pennsylvania, Ohio, West Virginia, and portions of Kentucky. Nearly all movement was halted, trapping people at home, at work, and on the road. Police, firemen, and countless volunteers provided emergency medical aid, and the National Guard helped maintain order. Businesses and many churches closed, football games were canceled or became farces, and most criminals even took a holiday.

With an ordinary snowstorm, the effects would generally diminish within 24–48 hours. Additionally, Superstorm 1950 occurred on a Friday and Saturday. Light traffic on Sunday (relative to weekdays) would ordinarily enable an even faster return to normal. But this was no ordinary snowstorm. As the next chapter details, the following week was anything but routine across the snowbound area.

4

DIG-OUT DAYS

"The work that is ahead is beyond any small group, beyond any single corps or body of men. It is everyone's job."

—PITTSBURGH MAYOR DAVID L. LAWRENCE, SUNDAY, NOVEMBER 26, 1950[1]

RESIDENTS IN SNOWBOUND AREAS AWOKE ON MONDAY, NOVEMBER 27, 1950, to typical winter weather. Skies were cloudy; many places, especially in Pennsylvania, had lingering flurries, but accumulating snow was done except downwind of Lake Erie, where only minor amounts would fall over the next few days. However, a simple look out the window made it obvious that the upcoming week was going to be wholly atypical. Nearly every road in western Pennsylvania, eastern Ohio, West Virginia, and northern Kentucky was buried in 1 to 3 feet of snow. Travel was nearly impossible. Phone operators, medical personnel, and law enforcement were stuck at work; travelers were stranded at hotels, churches, country farmhouses, and all manner of other places. Only a few long-distance trains were still running, and they were severely delayed. (Good luck reaching the station, too.)

The inability to travel really didn't matter for most people, though, as there wasn't anywhere to go. Nearly all major factories, including the colossal steel mills of Pittsburgh, were shuttered, idling hundreds of thousands of workers. Offices and downtown department stores were closed, schools and colleges had canceled all classes, and many newspapers could not publish. The region was completely paralyzed.

Despite highway workers' valiant efforts, Superstorm 1950 had overwhelmed them. It had destroyed most of their equipment by causing burnt clutches, engine seizures, or other mechanical failures; in several West Virginia counties, every single piece of snow removal equipment was busted. Many, whether working or broken, were stuck in snowdrifts, sometimes blocking streets. Worse yet, unusually cold November air lingered. Mother Nature was not going to provide assistance by melting the snow quickly.

Larger, national trends in transportation also delayed cities' reopening.[2] In the early twentieth century, cities gradually abandoned traditional rail-based systems and replaced them with bus systems. Personal automobiles also became wildly popular, and the rise of the automobile encouraged development in new areas far from the central city and existing mass transit. Residents in newly developed suburbs relied on their cars for transportation. However, automobiles performed miserably on snowy pavement. For those still using mass transit, buses had similar traction problems in snow. Thus, small snowfalls became disruptive events, and large snowfalls became disasters; no city had anywhere near the quantity or quality of equipment necessary to clear pavement in a timely manner. Additionally, the growing popularity of automobiles created more demand for downtown parking, and large piles of snow exacerbated the problem by reducing the supply of spaces. Motorists able to drive downtown often found themselves circling endlessly in search of parking, and this contributed to immense traffic jams late in the week after the storm.

Nonetheless, in the immediate aftermath of the storm, leaders and community members did not have time to ponder the chilly forecast or issues caused by the rise of the automobile. Their focus was on clearing the snow as quickly as possible. Many still had fresh memories of two world wars, and they approached this disaster with the same vigor and sense of duty. Using language that referred back to the Normandy landings, or D-Day, they called for thousands of citizens (meaning men and boys) to take up their shovels and fight the snowdrifts in dig-out days, or "D.O. Days." Although dozens would die, the army of volunteers successfully defeated the snow and helped normal life to return.

MONDAY, NOVEMBER 27

D.O. DAY

Although flurries were falling, conditions on Monday, November 27 were improving as the accumulating snow was over and the wind was dying down. This allowed weary road crews to make sustained progress. In cities, they focused on reopening major arteries and transit systems, while road crews outside of cities concentrated on major interurban roads.

Municipalities threw every available resource into digging out and made whatever sacrifices were necessary. In Lexington, Ky., the *Herald* described how garbage collection was delayed so workers could assist in the snow removal effort. In Clarksburg, W.Va., the *Telegram* noted that waste department employees were reassigned to help those with the street department. In Wheeling, W.Va., the *News-Register* reported on the city manager's successful appeal for help. A hundred extra men were added to the city's payroll to assist the snow-clearing effort. Likewise, municipal officials in both Butler and Pittsburgh, Pa., hired hundreds of temporary laborers, at rates ranging from $1.065 (Pittsburgh) to $1.50–$2.00 (Butler) an hour, to clear the streets. In Lima, Ohio, mayor Stanley A. Welker directed all Boy Scouts to "report with shovels" to aid in the task of street cleaning, according to the *Toledo Blade*. The *Charlestown (W.Va.) Gazette* noted that scouts in the local Buckskin Council of West Virginia were asked to clear snow away from fire hydrants.

Governors mobilized National Guard troops to assist road crews, laborers, and police officers. In Ohio, two thousand men from twenty different units helped in at least eleven cities throughout the eastern two-thirds of the state. Across the Ohio River in West Virginia, National Guard units delivered milk, cleared roads, and rescued a herd of twenty-five marooned dairy cows near Point Pleasant, as described by the *Charlestown (W. Va.) Gazette*. In Pennsylvania, the National Guard was most active in Pittsburgh, where guardsmen aided road-clearing efforts, and later in the week enforced a blockade of downtown. As conditions improved, the National Guard units in the three states were gradually released late in the week. Local newspapers reported that most in Cleveland were released by Thursday, November 30; in Youngstown, half were dismissed at midnight on Thursday night, with the remainder released forty-eight hours later; and in Pittsburgh, all units were demobilized at the end of Saturday as well. The governors of Ohio and Pennsylvania also declared legal holidays to permit banks in certain areas to close for the day. It's unclear how many actually took advantage of the holiday, but in Pennsylvania the order covered seventeen western counties. Pennsylvania's governor, James H. Duff, who was on vacation in Florida when the storm hit, had wisely flown back to Harrisburg on Saturday night. Governor Duff deserves credit for cutting his vacation short to follow what he said to the *Philadelphia Inquirer* was a "hunch." At the time, the storm was still raging and the impacts were unknown.

Governors in Pennsylvania and Ohio also supervised removal efforts on major roads. Both ordered traffic barricades at the state line between Pittsburgh and Cleveland to give crews a chance to clear roads. Unfortunately, these barricades caused traffic jams at East Liverpool and Youngstown, Ohio, later in the week as road conditions began to improve and interstate trucks piled up at the state line, as reported by the *Columbus Dispatch*.

Mayors jumped at the chance to show leadership, wisely recognizing that a slow or weak response could cost them their job.[3] The mayors of Pittsburgh and Cleveland stayed on the job nearly all week, and both vowed not to go home until the situation improved. It appears that at least one stayed true to his word; the *Plain Dealer* reported that

Cleveland's mayor did not go home until the night of November 30. In Charleston, W.Va., the *Gazette* observed that the mayor "personally supervised" the "giant clearing operation." He boasted that the streets would be "fixed by tomorrow [Tuesday]," an ambitious and likely impossible goal. Still, most major roads were reopened by Wednesday.

Most cities declared a state of emergency, which gave their leaders additional powers and access to funds. Youngstown, perhaps concerned about looting after the reports from Cleveland, issued a nighttime curfew, and taverns were not exempted. Police and firemen there were put on 12-hour shifts, and all leaves were canceled. The curfew was not lifted until Monday, December 4, according to the *Vindicator*. Extended work hours were mandated in other major cities as well.

Municipal governments borrowed any equipment they could get their hands on. Some came from local contractors and industry, while others came from other governments and the National Guard. Most local governments, for reasons including cost, storage, and maintenance requirements, did not own enough equipment to dig out from such a large storm. Even cities like Buffalo and Syracuse, both in New York, that receive large amounts of snow annually were perpetually short on equipment.[4] Perhaps such a mindset reflected memories of the Great Depression. Or perhaps the explanation is more tangible: 1950s snow-removal equipment was both more expensive and much less reliable than equipment today.

Because of the exceptional nature of the storm, community leaders and residents generally accepted that their governments had insufficient equipment. This was true with one major exception: Erie, Pa. Unlike the other cities affected, Erie averaged much more snow—around 90 inches each season, roughly double what Cleveland experiences and triple Pittsburgh's annual average. Thus, criticism rained down over the city's slow response. On both November 26 and 27, the *Erie Times* ran editorials blasting the city's response to the storm, calling it a "flop." The editorial board felt that the city's approach of waiting until after the storm "paralyzed" the community. The board continued:

We believe the city's plan of waiting until a severe storm hits and then—too late in this instance—renting bulldozers and similar equipment is a complete failure. Erie gets severe winter storms frequently, and it is high time the city acquires the modern equipment it needs.

In response, the mayor fired back by referring to such calls for more equipment as "ridiculous" and adding that the city "would go bankrupt if we tried to buy enough equipment to handle an emergency like this." He also noted, accurately, that Erie had done a better job of cleaning up than Cleveland and Pittsburgh, even if it had been slower than what critics wanted. Interestingly, as the week continued and Erie returned to normal while Cleveland and Pittsburgh remained paralyzed, the editorial board calmed down and praised the city's overall response and the work of the city employees, though it continued to repeat the call to buy additional equipment. Interestingly, almost three weeks after the storm, the city quietly decided to lease a snow loader for $300 per month. The city also had an option to apply the lease payments to the loader's $14,251 purchase price. In modern dollars that cost is approximately $150,000, which underscores how expensive such equipment was.

THE MYSTERY OF THE "DESTROYED" SNOW LOADER

Although the *Erie Times*'s editorial coverage of Erie's response to the snow became more favorable as the city returned to normal, a series of critical reports about a "destroyed" snow loader slung mud on everyone involved. On November 29, the *Times* ran a picture of a snow loader clearing snow at the Pennsylvania Railroad roundhouse. The loader had allegedly been "totally destroyed" in a city garage fire in August 1948 and sold at auction for $404, a fraction of its original $10,250 price. After restoring the loader for the cost of "two parts and a coat of paint," local contractor Jerry

Quirk claimed that he attempted to sell it to the city for $6,500 and was rebuffed. The editorial board, apparently eager to again criticize the city government, ran a harsh editorial the following day pointing out the foolishness of ignoring such an offer. However, elsewhere in the paper it was reported that an "angry" streets director, Tom McCarty, countered that (1) the city never actually owned the scrapped loader, and thus did not make the decision to scrap it, (2) it was old and inefficient and not worth the discounted price, and (3) although Quirk had contacted him about the loader, Quirk never submitted a bid when they had advertised for bids, meaning that he never formally offered to sell it back. No additional stories about the "destroyed" loader appeared, so it is unclear what the true story actually was. Perhaps this controversy is an example of the storm providing an opportunity for those with other agendas to advance them. (A similar thing happened in Scranton, Pa., in connection with the windstorm.)

Other cities also wrestled with whether to purchase snow-clearing equipment. In Youngstown, the mayor borrowed twenty machines costing $15,000 to $20,000 each (in 1950 dollars; modern costs are around $150,000 to $200,000 each) and told the *Vindicator* that purchasing even half that number should be considered "an extravagance." In Cleveland, the mayor acknowledged to the *Plain Dealer* that it might be sensible to buy a few things, but not units with a cost of $80,000, which also required the city to build new storage facilities.

With cities focused on reopening major arteries, some residents undertook collections to clear neighborhood streets out of fear that fire trucks could not reach fires. Several fires caused more damage than expected, since crews were hampered by poor road conditions. In Pittsburgh, the mayor encouraged residents to form "block parties" to prevent disaster, as proclaimed in the *Press*. His comments were underscored by a fire that destroyed a duplex in the Squirrel Hill neighborhood after fire equipment became stuck in snow en route. Ironically, residents of the street where the fire occurred had paid $54 to have their street cleared, but the trucks got stuck on Beechwood Boulevard, which had not been cleared. Residents in other Pittsburgh neighborhoods, as well as those in Cleveland, sometimes paid $1 per house to have contractors clear their streets. In Columbus, one neighborhood of 120 houses paid $0.50 per house to a neighbor with a bulldozer to clear sidewalks. Only three families declined to participate in the effort, according to the *Dispatch*.

Fire chiefs in many cities also warned residents to clear hydrants and use extra care with heating equipment. In Franklin, Pa., firemen themselves were summoned to clear hydrants, but this was an exception as residents were generally expected to do this. In some places Boy Scouts were specifically asked to aid this effort.

In the end, both public and private fire prevention efforts worked as few major fires occurred in the snowbound area. Sadly, despite these efforts there were seven fire-related deaths, including five children. Poor road conditions contributed to three of the deaths.[5] Nonetheless, the limited number of fire deaths and the lack of large fires were a sharp contrast with the deadly conflagrations that plagued the Deep South.

THE MIGHTY MILLS ARE CLOSED: EFFECTS ON INDUSTRY

The heavy snow shut down steel mills in major industrial cities and coal mines in West Virginia. In Pittsburgh, more than 250,000 industrial workers were idled, including more than 50,000 employees of U.S. Steel and its subsidiaries and 16,000 employees of Westinghouse. Experts quoted in the *Pittsburgh Press* estimated $100 million in production losses ($1 billion in modern dollars) and millions more in lost wages from the closures. U.S. Steel gradually reopened over the next several days, while the Westinghouse plant did not reopen until Thursday, November 30, because of both transportation problems and a lack of parking. Similar shutdowns affected hundreds of thousands of additional workers in Cleveland, Youngstown, and other major cities. In Akron, more than 100,000 workers were temporarily without work when the "Big 3" rubber plants closed. B.F. Goodrich managed to re-open on Tuesday, but Firestone and General Tire remained shuttered until Wednesday. Production and wage losses in Cleveland, Akron, and Youngstown were estimated at a combined $47 million (in 1950 dollars).

Although the roads were bad, thousands of workers in Louisville, Ky., were kept home on account of a gas shutdown. Fifty companies had previously agreed to stop using gas in extreme cold for reduced gas rates, and low gas supplies caused this plan to be implemented on Monday, November 27, according to the *Louisville Courier Journal*.

The snow rendered most coal mines inaccessible, and all coal mines beyond walking distance closed. On Monday, November 27, this meant that 30 percent of mines in western Virginia were closed, and nearly 100 percent of mines were closed in West Virginia, as reported in Roanoke and Charleston. The following day, about 40 percent of those near Charleston were still closed. Most mines did manage to reopen by Wednesday. Nonetheless, with the Thanksgiving holiday and sub-sequent storm, the *Bluefield Daily Telegraph* recorded that overall coal loadings for the week were down about 30 percent. In Pittsburgh, coal production dropped by nearly 1 million tons for the week, disrupt-ing income for more than forty-eight thousand miners, as reported in the *Press*.

Pennsylvania's and West Virginia's hunting seasons, which began the Monday after the storm, were also disrupted by the heavy snow. While a light layer of snow on the ground makes for easier tracking of prey, the thick layer was simply too deep to allow hunters to move easily—presuming they could reach their hunting camps at all. In West Virginia, the *Clarksburg Telegram* reported that overall deer kills dropped more than 60 percent from the previous year's numbers, while in some particularly rugged and/or isolated counties, the decline was greater than 80 percent. Further north, numerous hunters planning to travel to the mountains of northern Pennsylvania simply canceled their trips. The few who attempted to brave the elements had great difficulty, and area businesses counting on their patronage considered the start of the season a failure. One business that did do well was a planing mill near Titusville, Pa. Workers there created snowshoes by attaching hardware cloth to a plywood frame. (Hardware cloth is not cloth at all. It's a galvanized steel mesh typically used for applications such as animal cages and sifting.) On the first day alone they sold more than thirty pairs, and according to the *Oil City Blizzard*, on subsequent days the snowshoes sold out as quickly as they were made.

"SUICIDE BY EXERTION"

With workers idled, schoolkids home, and many streets snowbound, residents spent their time digging out. Unfortunately, modern consumer-grade snowblowers were not available until 1952, when Toro released the Snow Hound, so those cleaning up Superstorm 1950 had to shovel or push snow by hand. Snow shoveling is a very intense activity; it burns more calories per hour than anything except strenuous exercise. Furthermore, those shoveling (or walking long distances in the snow) are less likely to notice their bodies overheating, and it is easy to overexert oneself by focusing on completing the task rather than the body's danger signs. Thus, the deadliest hazard in snowbound areas was overexertion.

By Monday morning, more than forty-nine residents of Ohio had already died from the snow, with about half of those deaths resulting from shoveling snow.[6] Most of the remaining deaths, according to the tally in the *Columbus Dispatch*, were associated with heart attacks brought on by trying to walk through the snow or push stuck automobiles. It was a similar story in Pennsylvania, where more than twenty-three deaths occurred within the first few days after the storm, and also in West Virginia. Coroners, physicians, and newspaper editorial boards warned residents to use caution when shoveling, with an editorial in the *Columbus Dispatch* referring to shoveling deaths as "suicide by exertion." It's not clear whether these appeals had a positive effect. The death toll doubled through the week as additional heart attacks occurred.

Deaths from the snow were highly gendered and age-specific. Nearly all heart attacks afflicted men, and the majority were between 40 and 65 years of age. Men died while shoveling, after coming inside from shoveling, and from attempting to walk long distances through waist-deep snow. By Friday, more than 128 storm-related deaths had occurred in Pennsylvania and Ohio. According to AP reports, greater than 75 percent of those were associated with heart attacks or exertion. Only three deaths were women, and one of those may have survived had the ambulance been able to reach her, but stalled automobiles near Lima, Ohio, significantly slowed its progress, according to the *Toledo Blade*. Women mostly did not participate in snow shoveling. In fact, it was considered newsworthy that the Pittsburgh YWCA had its female residents help clear paths to its three area residences. The story was reported in the *Pittsburgh Press* under the headline "'Weaker Sex' Shows Strength in Emergency." The coverage termed it "sad" that "young men" were not doing the work, and quoted the resident director as saying it would be a "dateless" weekend. In the years following 1950, women have gained many more rights, and snowblowers are widely available. However, recent research examining snow-removal deaths from the early 2000s found that middle-aged men continue to make up the overwhelming majority of snow-removal deaths.[7]

Outside of those mentioned in the previous chapter, there were almost no fatal storm-related automobile crashes, perhaps because of the great difficulty in traveling at all. Three other snow-related accidental deaths occurred in West Virginia. One was when a grader being used as a snowplow overturned and crushed 58-year-old Price Snyder. Another was caused when a snow-packed chain failed to hold a coal car and it crushed 27-year-old Bernie Stover. And there was a tragic case where 12-year-old Sonya Lee Dailey sledded onto a roadway and into the path of an oncoming truck.

Although falling icicles created a hazard in many larger cities, they caused only one death, that of a steelworker in Pittsburgh, as announced in multiple news outlets. Snow and ice sliding off roofs was another potential hazard, but this was more of a problem for parked cars than people. In Oil City, Pa., for example, the *Blizzard* reported that two cars were damaged by snow sliding off St. Stephen's Catholic Church. The large stone structure has a steep south-facing roof high above and immediately adjacent to First Street.

Despite many people being stuck at home, there were no news reports of cabin fever. However, several days after the storm there was a road rage death. On Thursday, November 30, in the Oakland neighborhood of Pittsburgh, Arthur E. McCauley, 37, warned Carl Ramer, 51, to drive more carefully due to kids sledding in their neighborhood. A scuffle ensued and Ramer was knocked to the ground. Ramer then grabbed a screwdriver from his glove box and swung it at the much larger McCauley, fatally stabbing him in the head. (McCauley succumbed to his wounds two days later.) Charged with murder, Ramer, described by the *Pittsburgh Press* as "slightly built and graying," pleaded self-defense. In his three-day trial, which took place the following February, those testifying included the defendant, his 20-year-old son, and at least five children from the neighborhood. The youngest child was just 9 years old, and his testimony was permitted only after the judge asked if he understood the consequences of lying on the witness stand. (The boy replied, "Yes, sir, I'll go to hell—to the bad place.") After 5 hours of deliberation, a jury of nine men and three women acquitted Ramer.

LEGALLY EXCUSED: HOW THE STORM AFFECTED COURTS, DRAFTEES, AND BANKS

Courts were disrupted by the storm, with Pittsburgh courts adjourning at 2:30 p.m. on Friday, November 24. On Saturday, prothonotary (clerk of court) offices in Pennsylvania attempted to open, but the snow prevented most employees from traveling to work.[8] Not that it mattered; not a single court filing was made in Erie, Butler, or even much larger Pittsburgh. A typical Saturday in Pittsburgh had several hundred court filings.

Courts were expected to open on Monday, November 27, but with most judges, attorneys, staff, witnesses, and jurors unable to travel, they remained closed. Charleston, W.Va., and Cleveland courts closed for one day. Those in Columbus, Ohio, closed for two days, while Pittsburgh courts, hamstrung by the city's driving ban, stayed closed longer, and then only opened for emergency cases. Courts in Clarksburg, W.Va., were also closed, though three hardy jurors showed up for jury duty anyway, according to the *Telegram*. With little weekend crime, the widespread court closures did not have any serious effects.

Judges recognized the travel difficulties rural residents faced and were generally lenient in excusing them from jury duty and other court appearances. The military also took a relaxed stance, telling hundreds of Ohio, Pennsylvania, and West Virginia draftees scheduled for induction not to report.

Banks in 1950 were required to open on Monday by law, but as a result of road conditions the governor of Ohio declared a bank holiday, permitting them to remain closed. Most banks opened anyway, though local newspapers reported that they were short-staffed. In Pennsylvania, Governor Duff announced a bank holiday for seventeen counties in the western part of the state, roughly west of a line from Erie to Somerset. Nonetheless, banks in Erie and other northern parts of the holiday area managed to open anyway. In contrast, those in Pittsburgh remained closed. In an era without automated teller machines and credit cards, the extended closure triggered a citywide cash crunch, where a number of merchants and hotels ran out of cash or change. Although the bank holiday continued on Tuesday, five downtown banks opened with limited staff for "emergency" service—specifically for cashing checks and making change. The purpose of the bank holiday was not so much about safety concerns as it was to avoid loan defaults resulting from debtors being unable to travel to make payments. At least one bank ran a newspaper advertisement later in the week promising leniency for rural residents still unable to make payments because of poor road conditions. In 1950, there were no telephone or online instant payment options.

TUESDAY, NOVEMBER 28

THINGS BEGIN MOVING

Although cloudy and cold conditions still lingered (with pockets of light snow) region-wide, by Tuesday snow-clearing crews were making measurable progress. Government officials now reckoned with the biggest impediment to additional progress: abandoned cars. In all cities, abandoned cars were everywhere. They hampered traffic flow, blocked trolleys and buses, and filled valuable and limited parking spaces.

Cars today are much safer and more reliable than those in 1950.[9] At that time, stalling, especially in winter, was a major problem. Frequent stalling was caused by a variety of problems, including the use of carburetors instead of fuel injection, less-refined poorer-quality leaded gas, and shorter-lived spark plugs made of copper or steel instead of platinum.[10] Cars also handled more poorly in snow. They were generally rear-wheel drive, the suspension systems were of lower quality, and the tires were less able to grip the road. If the mechanical and handling issues weren't problematic enough, the streets themselves were more likely to be slushy or covered in snow, as snow-clearing equipment had all of the same mechanical and handling issues that cars had. These factors combined to create a cascading series of failures. Snow-covered streets and handling issues made it more likely that cars would slip and then stall. Motorists who were lucky enough to get moving again often had to stop again thanks to other stalled cars, repeating the cycle of slipping and stalling.

Repeated slipping and stalling increased the risk of complete mechanical failure; stalling and getting stuck in snow was hard on cars, and media reports were filled with stories about burned-out clutches, empty radiators, and broken springs and axles. With tire technology of lesser quality than today, tire chains were essential, but supplies often ran short. In Columbus, for example, tire chains ran out and new supplies did not arrive until December 1, according to the *Dispatch*. Of course, while all of the mechanical and handling issues were occurring, snow was falling, and the numerous stopped/stalled cars hampered road crews, leading to more snow collecting on the roads, and more problems for cars that were still operational.

Simply stopping at a red traffic light or stop sign could cause a car to stall or get stuck, so motorists sometimes ran them, creating additional hazards. This was especially common in the hilly cities of the South, like Asheville, N.C., but it was also a problem in the snowbound areas. For example, the *Columbus Dispatch* reported that motorists there had become so accustomed to crashing red lights during the storm that the police had to warn them to stop as the road conditions improved the following week.

During the storm, rapidly worsening conditions and mechanical failures caused many motorists to stop driving, regardless of where they were. Tens of thousands of cars were abandoned throughout the region, many of which blocked driving lanes. More than ten thousand cars were abandoned in Cleveland, and an estimated forty-five hundred were left in Columbus streets. Tow trucks were overwhelmed with calls for aid. In Pittsburgh, for example, the *Press* recounted that one company that normally handled 15 calls a day had more than 200 calls waiting, despite having nine trucks out. Another company had 354 calls for four trucks, while a third had 100 calls waiting but couldn't even get its own trucks out. Tow trucks had the same issues as other vehicles. In Cleveland, the *Plain Dealer* reported that more than thirty-four of eighty-four trucks operated by the Cleveland Automobile Club were disabled from burned out clutches, broken axles, and other mechanical problems.

Thousands of abandoned autos continued to litter the roads as the week wore on. Many more were operable but parked on curbs or in driving lanes since parking lots, side streets, and alleys were still snow clogged. In Cleveland, where more than four thousand cars were still snowbound on Wednesday, police began ticketing owners to encourage them to move their vehicles elsewhere. The *Plain Dealer* wryly noted that this tactic actually backfired, as many owners were more than willing to pay $8 in fines and towing charges to avoid digging out a stuck car themselves. Similar problems with clogged streets occurred in Pittsburgh, where city officials estimated that five thousand cars were still marooned one week after the storm began. Authorities pleaded for owners to move them, especially from streets with trolleys, but they did not resort to mass ticketing like in Cleveland. Instead, they simply towed cars down the street or around the nearest corner. A police spokesperson quoted in the *Press* dryly noted that if owners had trouble finding their cars, "it's their own fault for not having dug them out before this [time]."

Although abandoned cars were a disrupting factor, mass transit and intercity transit systems were able to resume limited service on Monday. By Tuesday, more routes were operational and area airports started to reopen. By Wednesday, nearly all mass transit between cities (airports, interurban buses, and trains) was close to normal. Within cities the majority of buses and trolleys were also running, though there were problems with delays and crowding as routes were limited and demand was much greater than normal.

Government services such as mail delivery and garbage collection were disrupted by the storm but gradually returned to normal during the week. Postal workers were aided by reduced mail volume, and delivery to many addresses could be skipped if mailboxes were obstructed by snow. Unsurprisingly, urban and letter delivery services resumed more quickly than rural and package delivery services, though all forms of delivery were normal, generally speaking, by the end of the week. Garbage collection in the hardest hit areas was delayed for up to week by impassible roads and a lack of manpower; crews were often reassigned to help clear streets. In Wheeling, W.Va., the *News-Register* reported delays in low-lying areas due to flooded roads and an unusually large volume of trash resulting from flooded basements.

THE EDUCATION SITUATION, PART 1: GRADE SCHOOLS
With rural roads across the region drifted shut and city streets and sidewalks helplessly clogged with snow, schools in the region had little choice but to extend the Thanksgiving vacation. Only those on the edges of the storm, such as in Toledo, Ohio, Louisville, Ky., or Harrisonburg,

Va., managed to open on Monday, November 27. Nonetheless, administrators were left scrambling as teacher absenteeism was much greater than normal. In Louisville and Jefferson County schools, for example, the *Courier Journal* disclosed that more than 190 teachers missed school. Most had spent Thanksgiving with relatives in snowbound areas and were unable to return by Monday.

City schools on the edge of the heaviest snow—meaning those in cities in western Ohio, Kentucky, and western Virginia—opened before rural, or county, schools. Drifting snow, especially in the flat, wide-open country of northwestern Ohio continued to cause problems for rural schools through midweek. According to the *Toledo Blade*, absence rates remained high, and some rural schools there opened Tuesday only to close again on Wednesday. Rural schools in far southwestern Virginia, where more snow had fallen, remained closed for the entire week, as reported by the *Roanoke Times*.

In paralyzed West Virginia, where school systems follow county boundaries, administrators had to contend with widely different conditions between urban and rural areas. Kanawha County schools, which included Charleston, reopened on Wednesday, November 29, but were unable to provide bus service, causing more than seventeen thousand students from outlying areas to miss class. A hundred miles northeast, a similar situation occurred in Clarksburg, where the attendance was described to the *Telegram* as "fair." In the southern part of the state, the school board of Mercer County (which includes the city of Bluefield) decided to shutter schools for the entire week rather than have more than six thousand bused students absent, as the *Bluefield Daily Telegraph* recounted.

In western Pennsylvania, schools in Erie, Butler, and Oil City reopened on Tuesday (Erie) or Wednesday (Butler, Oil City), but like those in other states, suburban and rural districts remained closed longer; in some cases, they closed for the entire week. Superintendents described attendance as "acceptable," and it gradually recovered as the week continued and road conditions in outlying areas improved.

Where the greatest amounts of snow fell—especially in major cities and in and near the West Virginia panhandle—students had extended vacations. In Pittsburgh, more than two hundred thousand students in city, Allegheny County, and parochial schools were left idle by the weather. While a few smaller systems in heavily urbanized Allegheny County opened late in the week, the majority of students had an extra week of Thanksgiving vacation. The story was similar in Cleveland, where two hundred thousand students were kept home early in the week. Like Pittsburgh, only those in small densely developed suburban districts resumed classes before Monday, December 4. Akron, Steubenville, Youngstown (all in Ohio), and Wheeling (W.Va.) schools closed for the entire week.[11]

While drifting snow was not a problem in the larger cities, parking and driving bans and deplorable sidewalk conditions were factors cited in decisions to cancel school. Buried sidewalks meant that students were forced to walk on streets narrowed by piles of snow and abandoned cars. Officials in Cleveland recognized that this problem threatened to spill into the first full week of December and nicely asked residents and businesses to clear walks as early as Friday, December 1, in anticipation of schools reopening on Monday, December 4. On December 2, their appeal in the *Plain Dealer* was more strongly worded, and on December 3, increasingly frustrated city leaders threatened to prosecute businessmen and others who failed to clear sidewalks, singling out gas stations in particular. While there is no record of any citations being issued, the mayor also ordered the fire department to place one man at every public and parochial school on the morning of December 4 to help crossing guards get children to school safely. Pittsburgh had similar problems with sidewalks. Additionally, schools there suffered from "heavy" damage amounting to "thousands of dollars" in anticipated repairs. The *Press* reported damage to roofs, gutters, cornices, and wall plaster (the latter caused by ice dams); however, the damage did not delay the opening of schools. Because most of the damage was to building exteriors, work was expected to be completed in a manner that would not disrupt classes.

Finally, schools had to adjust their schedules to make up the lost time, which was no small task due to the large number of lost days. In some cases, Christmas vacation was delayed. Many schools had planned to begin vacation after Wednesday, December 20, so a simple fix was to hold school on December 21 and 22. Additional days were made up by curtailing breaks and holidays in early 1951, or simply extending the end of the school year. In an exception, West Virginia schools were not required to make up the days at all—they were lost to "Acts of God" (quoted from the *Bluefield Telegraph*).

THE EDUCATION SITUATION, PART 2: COLLEGES AND UNIVERSITIES
Since colleges house many students in residence halls, they reopened more quickly than grade schools. Like grade schools, those on the fringes of the storm opened first. The University of Kentucky, located in Lexington, reopened on Monday, November 27, though absenteeism was common. The *Louisville Courier Journal* noted that school officials waived nonattendance penalties for snowbound students, urging them to take "no unnecessary risks." Bowling Green State University, about 25 miles south of Toledo, also opened on Monday. Attendance there was also below average, and even the university president was absent as a result of the weather. (The *Toledo Blade* noted that he was stranded at his summer home near Sandusky, Ohio.) Most other colleges and universities outside of the hardest hit areas opened around midweek.

Ohio State, in Columbus, canceled the first three days of classes after Thanksgiving break, reopening on Thursday, November 30. Perhaps the school should have reopened earlier, as idle students, still simmering over the football team's loss to Michigan on November 25, started a snowball fight on November 27 that turned into a five-hundred-person riot. Acts of minor vandalism occurred, including several broken windows and damage to cars with Michigan license plates. Despite making a very public arrest, the first responding law enforcement officer was unable to calm the crowd.[12] However, when university president Howard Bevis arrived and ordered the students to stand down, the crowd quickly dispersed. The five additional police cruisers supporting him surely helped encourage students to heed his command.

West Virginia University, in northern Morgantown, also reopened Thursday, November 30, after postponing class three times, according to accounts in the *Clarksburg Telegram*. Officials there contended with a roof collapse at Mountainlair, a recreation center built just two years earlier out of navy surplus materials.[13] About 75 miles northward in Pittsburgh, Duquesne University also managed to reopen on November 30, but other area colleges, including the much larger University of Pittsburgh, remained closed for the entire week, according to the *Press*. Leaders of Ohio University, in rural Athens, decided on Tuesday to close for the entire week, a sensible decision given the university's isolated location in hilly southeastern Ohio.

Unlike grade schools, colleges generally did not worry about making up closures of a day or two. Longer duration closures were made up by shortening the upcoming holiday break. For example, as described in the *Wheeling News-Register*, West Liberty State, located 11 miles northeast of Wheeling, W.Va., made up three days by adding two days of classes (Dec. 21 and 22) to the current term and adding one to the following term. The day added to the following term, January 6, was a Saturday, making the first week back a four-day week instead of three days.

OPEN FOR BUSINESS (SORT OF)

The effects of the storm on retail establishments were less severe than those on industry, and business returned to normal fairly quickly. Stores in areas with lower amounts of snow (e.g., Lexington, Toledo, and Erie) experienced reduced sales on Saturday, but they were able to open as usual on Monday, so the storm had little effect. (In 1950 stores were customarily closed on Sundays.) Stores in cities with moderate amounts of snow, such as Butler, Pa., and Columbus, Ohio, reopened on Tuesday as normal (Butler) or slightly later than normal (12:30 p.m. in Columbus), as indicated by local reports. In harder-hit Youngstown, stores wanted to reopen on Tuesday, but the *Vindicator* reported that they stayed closed until Wednesday at the mayor's request. In Pittsburgh, the five major downtown department stores stayed closed Monday and Tuesday but ran advertisements in local newspapers announcing plans to reopen for Wednesday. However, liquor stores that could open (about 60 percent) did "brisk" business, and army-navy stores recorded "great business," with galoshes being particularly popular, according to the *Pittsburgh Press*. While West Virginia stores generally did not close, with many people snowbound in rural areas, business was described as slow throughout the week. However, warmer weather and Christmas shopping brought out extremely heavy crowds the following weekend, as photographs from several cities showed the streets jammed with shoppers.

Department stores did not report any shortages of goods. Grocery supplies also had minimal disruption in most areas. Most food was delivered by rail in Cleveland, Pittsburgh, and other places in Pennsylvania and Ohio, and rail service was the least disrupted of any transportation method. Milk supplies in those states were generally sufficient as well. Dairy employees attempted to maintain a semblance of normal milk deliveries. Late in the week, large advertisements placed by dairy company leaders were seen in many newspapers thanking their employees (and patient customers). In some sections of Cleveland and Pittsburgh there were shortages of milk bottles, but enough customers responded to company pleas to return them that there were no serious problems.

Milk and food shortages were more serious in Kentucky and West Virginia. In Kentucky, poor road conditions left more than 75 percent of farmers near Louisville unable to bring milk to area dairies, causing a brief shortage as described by the *Courier Journal*. The storm also delayed resolution of a strike by Kroger truck drivers. The strike had caused seventy-five stores to be temporarily closed in Kentucky and southern Indiana. As a result of the poor road conditions, management and union representatives were unable to meet for several days, extending the length of the walkout. The two sides finally managed to resolve their disagreements on Friday, December 1, ending a fifteen-day strike. Stores were expected to reopen the following week, according to the *Courier Journal*.

In West Virginia, farmers could not bring their milk to market, triggering shortages. Problems were most acute in Wheeling, where milk supplies fell to 10 percent of normal, and dairies ran short of bottles. The *News-Register* also reported $94,000 in milk spoiled (almost $1 million today), in addition to whatever milk spoiled on farms and never made it to market. Severe restrictions were implemented to limit milk sales to those with children only, and purchasers had to sign a slip with their name and address. Egg supplies also ran critically low and were rationed to a half-dozen per customer. Although milk was not rationed in Charleston, supplies there also ran low, and the *Gazette* reported on how the National Guard cleared paths to farms so milk could be brought to area markets.

Like in the South, tire chains were hot sellers, with supplies in Erie and Columbus both running out, despite one Erie supplier telling the *Times* that he had a three-year supply prior to the storm. In Youngstown and Columbus, there were rumors of tire chains selling on the black market for $20 to $25, though no profiteering arrests were made.

POST-STORM RESCUES

As the week continued, rescuers were deployed to isolated groups, primarily in rural areas, who were short on supplies. Near Pittsburgh, the *Press* told the story about how three announcers and two audio engineers were marooned at the WKJF radio tower, located on a hill high above the city. After five days, the station's public relations director managed to break through the snowdrifts, bringing them some much needed food. The station had remained on air for the entire storm. In Wheeling, W.Va., rescuers also scaled a large hill to bring emergency food and supplies to four women snowbound in their home 5 miles north of the city. On the way back, the rescuers brought back thirty-five 5-gallon cans of milk from a nearby farm and dug out a State Road Commission plow that was buried in the snow. The plow was undamaged and returned to service almost immediately, according to the *News-Register*. Finally, on November 30 near Oil City, Pa., volunteer rescuers used snowshoes to walk across 8 miles of fields to deliver food to a woman and her seven children, aged 1 to 10. She had been stuck at home since Friday, November 24, when her husband left for his job in Erie. The *Oil City Blizzard* reported that he had been unable to make the 60-mile journey home due to poor road conditions.

Rescuers also aided those trapped away from home. In a widely told tale from West Virginia, the Charlie Bartrum family of seven, ranging in age from 8 to 37, had ignored warnings by rangers to leave Monongahela National Forest on Friday, November 24. Marooned by more than 40 inches of snow and running low on food, they forced their way into a closed-for-the-season fire tower and contacted officials on November 28. Finally, after an 8-hour ordeal, rangers using a bulldozer reached the party. Although they were down to their last food rations, the entire party was still in good health. Further north but still in West Virginia, on November 28 the National Guard attempted to rescue thirteen miners stuck outside of Morgantown and rumored to be low on food. When guardsmen arrived at the site, instead of thirteen miners they found a crew of three maintenance men. The *Clarksburg Telegram* said that the men were in good health, had ample food, and had no desire to leave their post.

While the story of the "trapped miners" was humorous, the rescue stories involving destitute families were heartbreaking. In the rural Oak Ridge Drive section of Charleston, W.Va., natural gas pipes had not yet been laid. By midweek, a group of twenty-five snowbound families had run out of coal and were burning chicken coops, outhouses, and any other wood they could find. Representatives from aid agencies pushed more than 4 tons of coal and $386 in food up a steep hill to prevent catastrophe. Pictures in the *Charleston (W.Va.) Gazette* depicted residents wearing dirty rags outside of shabby houses. In Greene County, Pa., a barking dog trying to get into a rural farmhouse captured the attention of gas company employees on November 29.

Breaking into the house, they discovered five children, ages 5 to 16, with no heat and little food. Their mother had been stuck in Washington, Pa., since the snowstorm hit. On November 30, in rural Hocking County, Ohio, a family of nine was rescued from a "ramshackle" house more than a half mile from the nearest road. Rescuers found them huddled under blankets in an upstairs bedroom, with the children crying and the women "near hysteria." There was no heat and only a half jar of jam left in the entire house, according to the *Cleveland Plain Dealer*. Several other heroic rescues occurred in West Virginia. Thankfully, in all cases rescuers managed to get to trapped residents in time.

Several dramatic rescues occurred thanks to fast, levelheaded actions. In Lexington, Ky., the *Herald* told how quick-thinking workers and the snow saved 13-year-old Paul Brumback from serious injury and possible death. Brumback was sanding floors in a house. When an electric light fell, broke, and set fire to the liquid he was using, he was immediately engulfed in flames. The workers simply tossed the burning young man into the nearest snowbank. Although his clothing was ruined, he suffered only minor burns. A heroic rescue also took place in Warren, Pa. When 4-year-old Karen Johnson fell through ice into the Allegheny River, her fellow 4-year-old playmate, Linda Irwin, grabbed her hand and held on for dear life. A passing motorist saw the girls and immediately sought aid. The *Butler Eagle* reported that both were rescued without incident.

THE HUMAN SNOWMAN

The most incredible rescue was that of snowman Edward M. Andras, 26, of Cleveland, Ohio, a military veteran of slight build.[14] When his car ran out of gas during the height of the storm on Saturday, November 25, Andras parked on a side street just a few hundred feet from Chester Avenue, a major thoroughfare on Cleveland's east side. Tired and out of cash, he decided to sleep in his car until his bank reopened on Monday (he reportedly had $2,800 on deposit), despite his being ill-prepared for such extreme weather conditions. Six full days later (December 1), a patrolman responding to a call about a body in an abandoned car discovered Andras. Barely conscious, highly disoriented (he thought it was November 26), and so cold that his legs had turned blue, Andras was rushed to Lakeside Hospital, where doctors expected to amputate his legs or feet. When asked how he passed the time in his car, Andras revealed that he spent most of the time sleeping and occasionally ate snow. Astonished medical personnel hypothesized that his body was in a semi-hibernation-like state. During the ordeal his weight had decreased a shocking 45 pounds, from 135 to 90. Miraculously, no amputations were needed, and he recovered and lived another 53 years. [15]

WEDNESDAY, NOVEMBER 29, AND BEYOND
NEW PROBLEMS DEVELOP

Factories, large downtown stores, and other businesses began opening more widely by midweek. However, with large piles of snow everywhere, many cities had severe traffic jams—complete gridlock in the worst cases—from a lack of parking. Youngstown, Ohio, was one such city. Daily traffic jams there occurred from November 29 to December 2. The *Vindicator* noted that these were exacerbated because street crews closed lanes of busy streets and bridges to remove snow during rush hours.

Officials in all cities struggled to balance the need for reopening with difficulties caused by insufficient parking, abandoned cars, and poor road conditions (especially side streets). To reduce the number of vehicles, they exhorted people to use mass transit. Unfortunately, their requests were heeded by too many citizens. Transit systems in several cities were overwhelmed, causing people to have to squeeze on board like "sardines" (Pittsburgh) or wait hours for buses and trolleys (Cleveland and Youngstown). Transit systems were also slowed by heavy auto traffic and numerous abandoned cars blocking roads. In Youngstown (and likely elsewhere), some generous motorists offered rides to those stuck at bus stops.

In Canton, Ohio, a workers' strike meant that mass transit was not an option. Gridlock resulted, and the *Cleveland Plain Dealer* reported that some motorists took 30 minutes to simply cross an intersection. Nearby in Akron, traffic was snarled by thousands of trucks passing through town after Pennsylvania officials finally cleared the Pennsylvania Turnpike. In 1950, Ohio's turnpike was still in the initial planning stages, having been delayed by state officials caught up in funding fights and numerous other disputes.[16] There was simply no way for trucks to bypass Akron.

To clear out city centers faster, authorities in all major cities restricted or banned parking in downtown areas. Pittsburgh and Cleveland banned all downtown parking. More limited bans were issued in Butler and Erie, Pa., Youngstown, Ohio, and Wheeling, W.Va. These bans either limited overnight parking in certain areas or were for select streets to allow road crews unfettered access.

Parking bans and driving restrictions, which only the larger cities implemented, helped cities clear snow effectively, but caused other problems elsewhere. In Youngstown, the *Vindicator* told how downtown parking was banned and city workers set up barricades to block nonessential traffic from the core beginning at 5:00 a.m. Wednesday, November 29. Unfortunately, traffic at the barricades became so jammed that "essential" traffic was also blocked, and the barricades were taken down a mere 5 hours later. Cleveland also limited downtown traffic to "essential" activities only but did not use barricades to enforce it. As might be expected, many motorists considered themselves essential, so a severe traffic jam occurred on the twenty-ninth. Another traffic jam occurred the following day, but it was less severe. This was because city leaders had banned beverage trucks from blocking streets for deliveries for the rest of the week, according to the *Plain Dealer*. Officials there and in Columbus asked businesses to stagger their opening and closing times in an effort to reduce rush hour traffic. It's unclear how widely such requests were followed or if any benefit was realized. Road conditions were improving anyway as road crews made progress and abandoned autos were moved.

On Wednesday, November 29, Pittsburgh attempted to open up downtown (also referred to as the "Golden Triangle") to traffic. Unprecedented traffic gridlock followed. Although the mayor blamed retail stores and kids "yelling to see Santa Claus," the gridlock was actually caused by a lack of available parking. Parking was banned on some streets, and many off-street parking lots and spaces were still filled with snow. Motorists found themselves driving in circles looking for spaces, which further increased traffic volume.[17] Others simply double-parked, blocking trolleys and further restricting the number of travel lanes.

The city responded by setting up a complete blockade of the Triangle the next morning. Beginning at 7:00 a.m. Thursday, November 30, National Guardsmen manned forty-three checkpoints. The city issued approximately fifteen hundred consecutive numbered orange cards to essential workers and those unable to use public transit. Only motorists possessing a card or other strong justification to go downtown were permitted through the blockade. With more than 75 percent of motorists turned away, traffic backups at the checkpoints developed, though they were less severe than those at the barricades in Youngstown. Nonetheless, the blockade was highly successful. Rapid clearing of the Triangle resulted, and the blockade was lifted the following weekend. Although on-street parking was still prohibited downtown, nearly all off-street parking lots were now cleared, according to the *Press* report.

Some of the worst traffic jams occurred on Saturday, December 2, a full week after the storm. This may seem odd as it was a weekend day, crews had made great progress, and warm temperatures were melting the remaining snow. However, three other factors caused the jams. First, warming temperatures and continued progress from road crews meant that residents of rural areas could finally leave their homes and travel into town for supplies. Second, with Christmas just three weeks away, people were eager to shop. Recall, too, that residents had been unable to shop the previous weekend as a result of the snow. Finally, all major retail stores were located in city centers in 1950; the outmigration to suburban shopping plazas and malls had not yet begun.[18] These factors all combined to create terrible traffic jams in the region, most severe in secondary cities.

Two of the worst jams occurred in secondary cities in West Virginia. In Clarksburg, the *Telegram* claimed that "the worst traffic jam in the history of Clarksburg" occurred on Saturday, December 2. With temperatures there a balmy 60°F, the deep snow on many streets turned to slush, and many automobiles became stuck in ruts. Police blamed motorists for ignoring requests to drive with tire chains. In Bluefield, W.Va., the *Daily Telegraph* described traffic as being so heavy that patrolmen turned off traffic signals and directed traffic themselves to help it move more efficiently. In Butler, Pa., the story was similar. Slushy road conditions and hordes of shoppers and hunters created such gridlock that all available police officers (twenty-two of twenty-four) were ordered to direct traffic. They were not relieved of duty until after midnight, according to the *Eagle*.

FIGHTING SNOW WITH FIRE

Cities in the region were so choked with snow that road crews ran out of places to put it. In an era before clean water standards, many municipalities simply dumped the snow into the nearest body of water. In Pittsburgh, the *Press* printed a front page photo of a cavalcade of trucks dumping snow into the Monongahela River.

In Youngstown, Ohio, the *Vindicator* described how the Pittsburgh and Lake Erie Railroad used summertime weed burners to help clear the tracks. Instead of burning tall grass and weeds along the side of the tracks like usual, they melted snowdrifts adjacent to the tracks.

The most interesting snow disposal experiment transpired in Wheeling, W.Va. Wilbur Everett, a local bottled gas dealer, offered the use of his homemade flamethrower for melting snow. Though there were concerns about the potential for a fire, the city manager gave it the OK and told the *News-Register* that the "city had everything to gain and nothing to lose." The highly publicized experiment, reports of which ran nationally through the Associated Press, turned out to be a complete failure. The *News-Register* reported that the flamethrower only melted enough area "to cover a newspaper." An area physics professor belittled the experiment, saying, "The whole idea is silly." He explained that the amount of energy needed to melt even a small amount of ice was much greater than the capacity of the specially equipped fuel truck.

WATER, WATER, EVERYWHERE!

As the week continued, ice dams developed on many buildings and caused extensive damage. An ice dam is caused when heat from inside a structure melts snow on the roof, and the meltwater freezes over unheated eaves or gutters at the roof's edge. With nowhere to go, additional meltwater backs up under the shingles and enters the building, damaging interior ceilings, walls, and furnishings.

Ice dams were widespread following the storm (though most often discussed in Pennsylvania newspapers), and much more significant than roof collapses in terms of numbers of houses affected. Buildings in 1950 were much less insulated than modern structures. Worse yet, few homeowner's insurance policies provided coverage. "All perils" insurance did not exist, and insurance companies felt that water damages were unpredictable, so only a few, very expensive, policies covered such damage. Most residents, thus, had to pay out of pocket for damage from ice dams. Issues of limited insurance coverage also caused headaches for homeowners affected by the windstorm in the Northeast.

While most flooding associated with the storm occurred in other regions, rapid snowmelt in early December caused flooding problems in the snowbound areas. Luckily, autumn had been dry and the ground was unfrozen, so most meltwater soaked into the ground. Furthermore,

area reservoirs had excess capacity, which also reduced flood potential. The *Vindicator* reported that Youngstown reservoirs still had 50
percent of capacity available after the snowmelt, while the Tionesta
Dam, upstream of Oil City, Pa., was able to limit the flow of water
down the Allegheny, preventing major damage to towns immediately
downstream.

Nonetheless, given the sheer amount of water, limited flooding
did occur. Basements flooded in Erie and Oil City, Pa., and in Ohio,
around 150 residents in Eastlake (about 20 miles east of Cleveland) had
to evacuate the raging Chagrin River, according to the *Plain Dealer*.
The capacity of higher-order rivers—primarily the Ohio River—was
also exceeded. Pittsburgh experienced a crest of 3.5 feet above flood
stage on December 4. However, at this time nearly all snow had melted
and the weather was cold again, so the *Press* reported that no major
damage occurred. In Wheeling, W.Va., rapidly rising water on Sunday,
December 3, prompted more than 175 families to evacuate low-lying
Wheeling Island. Hydrologists quoted in the *News-Register* expected
the Ohio River to crest 9 feet above flood stage. But colder weather
halted the snowmelt, and the final crest was 40.7 feet, an amount less
than 5 feet above flood stage. While basements were inundated, relieved officials characterized the flood as a "minor annoyance." The
News-Register did use the occasion as an opportunity to advocate for
a flood wall. The editors noted that a flood wall in Huntington, W.Va.,
had prevented major flooding there and criticized the government for
"sending eight and nine-figure handouts" to Europe instead of protecting Wheeling.

While the flooding in Wheeling was minor, two flood deaths occurred near Roanoke, W.Va., about 35 miles south of Clarksburg. According to the *Clarksburg Telegram*, 4-year-old Johnny Fisher fell off a
swinging bridge into the raging West Fork River. His brother, Ferrell, 11,
attempted to save him, but he was also pulled into the torrent. Neither
was seen again. These children were the final two storm victims. Their
deaths occurred ten days after the first storm victims, Robert Nilles and
Ethel McPherson, died in car crashes on Thanksgiving night.

PAYING FOR THE STORM

While warm weather helped melt the snow relatively quickly, bills incurred for snow removal did not disappear as easily. Municipalities scrambled to cover the costs, often taking money from the street funds. In West Virginia and Kentucky, where the cities are mostly small in size, costs were in the thousands of dollars. Costs were estimated at $5,000 in Bluefield, $9,000 in Clarksburg, $2,934.80 in Charleston, and $1,000 per day in Lexington (probably around $5,000 in total). Costs were similar in Erie and Oil City, Pa., where the *Oil City Blizzard* noted that the storm had also caused a sharp reduction in two streams of revenue for the streets department: parking meters and parking tickets.

In larger cities in Ohio and Pennsylvania, costs exceeded tens or even hundreds of thousands of dollars. Youngstown spent around $60,000, Columbus's tab was about $100,000, and Pittsburgh's bill was nearly $500,000 dollars. In modern dollars, these amounts are approximately ten times as much. Pittsburgh's cost is only for the city government and does not include an additional $525,000 spent by other boroughs, townships, and the state within Allegheny County. In nearby Wheeling, W. Va., where leaders spent approximately $1,000 per day for snow removal, the recently approved 1950–1951 budget had no provision for any funding at all, leading the city to request donations. The fund drive was a spectacular failure. The *News-Register* reported that only three donations were received, and two came from businesses eager to reopen.

POSTSCRIPT

CONTEMPORARY REFLECTIONS ON THE STORM'S IMPACTS

By the first full week of December most of the snow was gone, and life had returned to normal for most people. Children were back in school, industrial output was rebounding, and area residents resumed their Christmas preparations. Although the storm had caused great inconvenience, widespread property damage, and dozens of deaths, any lingering effects were mostly borne by individuals and not society as a whole.

Generally, newspaper editorials, a barometer of local sentiment, were favorable. Most newspapers recognized the extraordinary nature of Superstorm 1950 and commended government, utilities, and citizens for their hard work. Many newspapers also praised the efforts of police, fire, and public works employees and lauded the community spirit shown by individuals in clearing streets and helping neighbors. Criticisms were mostly constructive and suggested better coordination, budgeting, or purchases of equipment.

Several editorials reflected the national mood by connecting the disaster with a potential nuclear attack. A common concern was that difficulties dealing with the storm demonstrated that many places were ill-prepared for a nuclear attack. The governor of Ohio also expressed a similar concern. Although not widely reported, some civil defense groups used the storm as an opportunity to review their response. In Cleveland, for example, the local preparedness committee created a subcommittee specifically to learn lessons from the storm response for application in the case of a nuclear attack, as described by the *Plain Dealer*.

Newspapers also meditated on how routine life had changed since past major storms. In Bluefield, W.Va., the editor of the *Daily Telegraph* noted that while daily life had improved vastly since the last major storm in 1913, many more people were stuck on the roads due to the rise and growth in automobiles and highways, a rare acknowledgment of how these changes had increased vulnerability. Furthermore, while automatic heating devices and refrigerators were certainly better than older methods of heating and cooling, electricity was required. Thus, if power was lost, people suffered and food spoiled. The *Columbus Dispatch* echoed those observations. The *Cleveland Plain Dealer* ran an editorial cartoon depicting a "good old fashioned snowstorm" striking down a solitary man with his "aloof, push-button living," giving him a "deserved comeuppance." In Oil City, Pa., *Blizzard* editors discussed how fallen trees in the past did not tear down wires, and people could just cut them down if they fell. In contrast, fallen trees now cut off electricity, causing inconvenience, and someone else had to be hired to dispose of the tree because people could not deal with fallen trees on their own anymore. The paper continued by comparing trees to kids—when well-behaved, they are a pleasure to have around, but when troublesome, they could be "very troublesome, indeed." Finally, in the *Cleveland Plain Dealer*, an economics professor reflected on how the storm slowed down the pace of life:

> Most Americans tend to live intensively. They work hard, move rapidly, and want results in a hurry. The snowstorm stopped this rapid pace.

Although the storm was disruptive, it also brought great joy to some, especially children. Many cities maintained sledding courses or allocated certain streets for this purpose. Newspapers everywhere showed many children enjoying the snow. The *Bluefield Daily Telegraph* described the sledding as follows:

> The old weather prediction: snow, followed by little boys with sleds; splattered the nail on the head yesterday. Bluefield kids took to the snow like tassels on a toboggan.

The joy children experience with a fresh snow is universal. Although kids in Altoona, Pa., also enjoyed sledding after Superstorm 1950 hit the area, the large amounts of ice that preceded the snow caused a blackout and crippled the city for more than a week. The incredible ice storm that affected west central Pennsylvania is the subject of the next chapter.

5

AN ICY BLACKOUT

"In my more than 40 years in the utility business, I have never seen or heard of a storm that caused such widespread destruction to electric facilities."
—D. W. JARDINE, PRESIDENT OF PENELEC, ON DAMAGE AROUND ALTOONA, PA.[1]

UP TO NOW WE HAVE DISCUSSED THE EFFECTS AND RECOVERY FROM record-setting snow in the eastern Ohio Valley. East of this area, in the region depicted in figure 7 and I describe as west central Pennsylvania, a different type of weather occurred.[2] This region is anchored by Altoona, which had a 1950 population of 77,177, and Clearfield, with a 1950 population of 9,357. Altoona, Clearfield, and most other cities and boroughs are at the bottom of narrow stream valleys, while nearby tree-covered ridges and mountains are hundreds of feet higher. Cold air below freezing often gets trapped in the valleys while warmer air streams overhead along the ridges. This setup is perfect for sleet and freezing rain and makes the region prone to ice storms.

Normally the colder air in the valleys warms quickly, and usually the precipitation is not heavy. But as Superstorm 1950 crept across the region, the low level cold air did not warm. Worse yet, the intensity of the storm caused a prolonged period of heavy precipitation. The result was one of the most destructive ice storms ever in the United States.

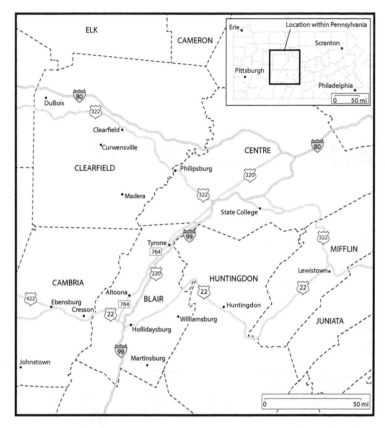

FIGURE 7. Locator map showing counties and cities around Altoona and Clearfield, where the heaviest icing occurred in 1950. (Interstate highways did not exist in 1950. They are shown for orientation purposes only.)

A MIND-BOGGLING AMOUNT OF ICE

Prior to the storm, weather conditions across the area were cool but typical for late November. Daily high temperatures were in the thirties and lows were in the twenties. During the week before the storm, light snow or rain fell in most places, but this was more of a nuisance than a problem.

After a chilly Thanksgiving with more flurries, Friday, November 24, was considerably warmer, though windy. Temperatures steadily rose from midnight until 3:00 p.m., when they peaked at 46°F in Altoona. A strong cold front passed through afterward, and in 1 hour the temperature plummeted to 30°F as westerly winds brought in much colder air. Temperatures continued to decrease as night fell.

Precipitation began falling during the evening. Even though it was more than cold enough for snow in Altoona with a temperature of 20°F, sleet began falling at 7:30 p.m. Several miles above the ground it was snowing, but as the snow fell it entered a warm layer of air that melted it into rain. Below that, the very cold air near the ground refroze the rain into ice pellets, or sleet. (Once a snowflake melts into a raindrop, it doesn't change back into a snowflake, regardless of temperature.)

The warm air high above the ground was coming from the Atlantic Ocean, and as Superstorm 1950 strengthened, this "conveyor belt" of warmth and moisture grew stronger. The air rose so violently that occasional lightning and thunder occurred, despite subfreezing temperatures. While some of this warmer air did work its way downward, a very thin layer of air near the ground remained below freezing. The depth was not enough to freeze the raindrops into sleet. Instead, the rain fell as liquid droplets and then froze on contact with objects near the ground—a condition known as freezing rain. Freezing rain was observed during the evening in Altoona and nearby Hollidaysburg. Further north in Clearfield, the freezing rain began accumulating heavily on power lines and trees around midnight, according to the *Progress*.

Freezing rain is the most destructive form of winter precipitation. Since it falls as liquid rain, it "sticks" and bonds as it freezes on contact with trees, power lines, and other objects with temperatures below freezing. Water is heavy: 1 gallon weighs more than 8 pounds. This means that relatively small accumulations of ice, on the order of 1 inch, can cause significant destruction.[3]

After midnight, the cold front that passed though on Friday reversed course and began drifting westward as a warm front. But the tree-covered ridges blocked the warm air from penetrating into the deep valleys of the region, and temperatures remained stuck below the critical 32°F value. Copious amounts of freezing rain fell from the rapidly intensifying storm—about 1 inch every 3 hours. These were much greater than with a typical ice storm. The freezing rain continued all night and into the following morning, when it finally ended around 10:00 a.m., according to the *Altoona Mirror*. At this time, the cold-turned-warm front reverted to a cold front and swept eastward once again, cooling the entire atmosphere sufficiently to change the precipitation to a heavy wet snow. Area residents were still not in the clear. The heavy wet snow rapidly accumulated at a rate of about 1 inch per hour. The additional weight caused further destruction as already overburdened trees and power lines were pushed past the breaking point. Significant accumulating snow finally ended in the afternoon.

The final precipitation totals were mind-boggling. Never had anyone in the area experienced so much freezing rain, and the ice accumulated in a very short period. As someone who wrote a dissertation on the impacts of ice storms, I am of the opinion that this was one of the most intense ice storms ever to affect the United States.

Nearly every reporting station in Blair and Clearfield Counties received 4 inches or more of liquid-equivalent precipitation. (The term *liquid-equivalent precipitation* refers to the depth of water obtained when all snow and ice is melted and combined with any rain that falls.) Such amounts are unprecedented for such a short period during a winter storm. In Clearfield County, Curwensville, Madera, and Phillipsburg received anywhere from 4.0 to 4.4 inches. In Blair County, Tyrone received almost 5 inches of rain. This was actually ordinary rain, not freezing rain, and it caused major flooding. Elsewhere in Blair County, Hollidaysburg and Altoona received around 4.5 inches of mixed precipitation.

It is unknown exactly how much freezing rain fell because of insufficient detail in hourly weather observations. However, Weather Bureau summaries and local news accounts indicate that a reasonable guess is anywhere between 2.5 and 3.5 inches, with isolated amounts of around 4 inches. Determining the exact amount of snowfall is difficult for similar reasons. It appears that around 6 inches of snow accumulated in Altoona, and Clearfield County received anywhere from 6 to 10 inches of snow.

Regardless of the amount, the bulk of the precipitation fell as freezing rain. Freezing rain accumulations of 1 inch or more typically cause catastrophic damage to trees and power infrastructure. The results were well beyond "catastrophic," suggesting that amounts were much greater than 1 inch. The destruction was so complete that most areas would be without electricity for an entire week.

THE ANNIHILATION OF THE ELECTRIC GRID

The environmental destruction from the ice was total and complete. Nearly every tree was severely damaged or destroyed. The Clearfield Shade Tree Commission estimated that 90 percent of all trees were damaged or destroyed, while in Altoona a full 100 percent of trees were considered damaged. Damage outside of the cities was equally severe. The Weather Bureau's climatological report simply noted that the rural areas were "laid bare of trees."[4]

The power grid was decimated from the ice weighing down lines and from trees falling on lines. In most of Clearfield County, outages lasted for days. In Blair County, conditions were worse: the ice completely destroyed the electric grid. While touring the area nearly two weeks after the storm, D. W. Jardine, the president of Penelec (the electric utility), was quoted by the *Altoona Tribune* as saying that he had never "seen or heard of a storm that caused such widespread destruction of electric facilities" in his 40 years of experience.[5] Widespread power failures in Altoona began around 3:00 a.m. on Saturday, November 25, with a complete blackout starting between 6:00 and 7:00 a.m., as described in the *Mirror*.

The magnitude of the destruction was incredible and cannot be overstated: not only were ordinary circuits and lines to individual homes and businesses brought down, but nearly all high-voltage transmission lines in the region were destroyed. In Clearfield, two of the three main transmission lines that served the borough were knocked out of service and four of the five distribution circuits were damaged. Five miles southwest in Curwensville, two of four feeders were damaged. In Altoona, 35 miles south of Clearfield, every single transmission line—six in total—was rendered inoperable by the ice; most were completely destroyed. The Altoona facts came from sources outside of Altoona, as the local newspapers were unable to publish regular, full editions until the Tuesday after the storm. (They did publish some limited token editions for their archives.)

The recovery time was unusually long. On November 27, two days after the storm, half of the borough of Clearfield lacked power. While most power was restored by November 29, the third feeder line and final repairs were not completed until around December 1, nearly one week after the storm, as reported by the *Progress*. Curwensville and rural areas outside of Clearfield were without power for much of the week as well. Persistent below-freezing temperatures and additional accumulating snow on November 28–29 further delayed the power restoration efforts.

In Altoona and the southern half of Blair County, recovery took even longer because of the epic scale of the destruction. The forested ridges were completely stripped of trees, with Brush Mountain among the worst affected locations. Not only did tree limbs continue to fall for days after the storm ended, but entire trees continued to fall for several days as the ice continued to weigh them down. Two transmission lines were restored on the day after the storm (Sunday), but falling trees knocked them back out of service on Monday. Frustrated workers would restore one portion of a line, only to have it collapse behind them as they moved on to the next segment. Penelec finally revived one 5,000-kilowatt transmission line on Monday, but this amount was insufficient for almost any activity (Penelec called it "shoestring service") and unreliable as additional failures continued to interrupt service. Thus, the Pennsylvania Railroad's (PRR) internal generating plant on the south side of Altoona was used to provide emergency service to three area hospitals, three bakeries, and two dairies. With falling limbs and trees continuing to disrupt service on both Monday and Tuesday, the PRR was the only consistent supplier of electricity in Altoona early in the week.

Hundreds of repair workers streamed in the week after the storm—three hundred on Sunday, a hundred more on Monday, and another hundred on Tuesday, as chronicled by the *Altoona Tribune*. Workers were housed in PRR sleeping cars and fed at the Jaffa Mosque (a local arena now called the Jaffa Shrine Center), where more than sixty-five hundred meals were served in the nine-day period from November 27 through December 5. To serve so many meals, the employees of the Jaffa Mosque put in a heroic effort. According to the *Altoona Tribune*, Simon Ebersole, steward of the Jaffa Mosque, asked for relief on December 5. He and his crew of twenty people were fatigued after averaging only 3 hours of sleep per night because of the heavy workload. Although they offered to continue serving if needed, Penelec thanked them and arranged for the Penn Alto Hotel to take over meal service for the dwindling number of repair workers.

Additional workers arrived after power was restored to Lewistown, Pa. (also shown in fig. 7), but persistent temperatures below freezing meant that the ice continued to weigh down the lines and break tree limbs. The *Altoona Mirror* described how desperate Penelec workers experimented by short-circuiting portions of the lines and using the heat generated to melt the ice, hoping to help relieve the strain. The experimentation was a success. The *Tribune* reported that it worked nearly everywhere except the highest ridges, like Brush Mountain, which were colder from the higher elevation. (Brush Mountain is approximately 1,400 feet higher than the city of Altoona. Given the elevation difference, temperatures there should, on average, be five degrees colder than those in Altoona.)

By Thursday, a full five days after power was lost, Penelec had restored a second 5,000-kilowatt line from Cresson, directly west of

Altoona, and was testing a third line from Tyrone. With two lines active and reliability increasing, the PRR station was no longer needed. Although there was insufficient power for the city's normal needs, it was enough for the moment. Around 75 percent of customers were still off grid, the result of failures in distribution circuits and connections to buildings. The lucky 25 percent were asked to conserve anyway.

Temperatures finally inched above freezing, to 35°F, on Thursday, November 30. Although the warmer weather help loosen accumulated ice, when chunks of ice fell, the lines violently snapped back and forth, causing new failures. This process further delayed restoration and testing of a sorely needed 25,000-kilowatt line from Williamsburg (east of Altoona) to the city, as reported by the *Tribune*. Additionally, failures from falling ice became so numerous that the PRR generating station was pressed back into service for a brief time.

Despite these issues, November 30 was the first of ten consecutive days where highs exceeded freezing, and the pace of repairs accelerated as other components, like transformers and switches, became ice-free. More workers arrived from elsewhere as power was restored in other cities and states (such as Vermont) affected by the storm, further helping matters. The worst of the disaster was over at precisely 6:25 p.m. on Friday, December 1, when the 25,000-kilowatt line was reenergized and all conservation restrictions were lifted. For the first time in nearly a week, electric lights shone on the streets of Altoona. Restrictions were also lifted because two lower voltage lines from Tyrone were also restored on Friday, helping increase reliability.

Although the thawing conditions caused additional failures in the region, restoration continued at an accelerated pace as the calendar flipped to December. Rural electric cooperative companies restored power to nearly all customers by the end of the first weekend of the month (December 3), according to reports from both Blair and Clearfield Counties. In Altoona, where all major transmission lines were now in operation, crews fanned out to restore the distribution network within the city. So many repairs were needed that it was several more days before most individual customers had power. But by December 4, nine days after the storm, power was restored throughout downtown, and by December 5, a full ten days after the storm, the majority of customers had power once again. Two days later, the *Tribune* listed that the number of customers with power had exceeded 85 percent.

At this point, however, progress slowed. The remaining 10–15 percent of customers without power, representing around four to five thousand households and businesses, were in areas where the electric grid was so devastated that it had to be completely rebuilt, a very time-consuming process. In one neighborhood a flood had occurred, toppling poles by their bases and ruining all equipment such as transformers and switches. (This flooding also delayed construction of a new sewage plant by two weeks, according to the *Mirror*.) In another area a 10-mile segment of line was damaged beyond repair and had to be completely restrung, a process that took the better part of a week. Finally, heavy rain and wind gusts to 55 mph in the valleys (and probably higher on the ridges) on Thursday, December 7, caused new problems and reversed some earlier gains, as reported by the *Tribune*. Luckily the setbacks were temporary, as improved weather afterward allowed progress to resume. On Monday, December 11, Penelec declared the emergency over and began sending home some of the additional workers. Finally, on December 14, for the last time the *Tribune* provided an update on repairs. Any remaining repairs were so few in number that they weren't newsworthy. December 14 was nearly three weeks after the storm.

Utility poles carry more than just electrical power lines, and phone service was also severely disrupted by the storm. Like electric service, phone service was restored more quickly in Clearfield County, where the *Progress* described how 1,200 customers in Clearfield Borough still lacked service on November 28. The following day, only 721 still lacked service. (Several hundred customers in other boroughs in Clearfield County did not have service either.) Meanwhile, in Altoona more than 5,000 customers lost service initially, and about half of those still lacked service six days later (December 1), as recounted in the *Tribune*. However, the thaw and additional progress meant that phone service was considered back to normal by the following Monday (December 4) in Altoona.

Phone service was most severely affected in the Morrison Cove area, an isolated valley southeast of Altoona. Here, the phone system had to be entirely rebuilt, and it was a slow process. Two and a half weeks after the storm, phone service was still unavailable for 40 percent of customers. A week later, phones still were not working for about half of those without service the previous week. Most without service were along the southwestern and southeastern edges of the valley. At that point, Bell Telephone warned customers that full restoration was unlikely before Christmas, one full month after the storm.

Like Penelec, Bell Telephone threw all of the resources it could spare at the problem. Initial reports listed an additional 321 workers, with 107 of them (along with fifty trucks) brought in via railroad from Pittsburgh due to snow-clogged roads. The ranks of telephone repairmen eventually grew to more than 400. Workers were housed and fed at the Penn Alto Hotel. A preliminary estimate from Bell Telephone projected losses of around $1,250,000, or about $12.5 million today. Repairs may have been completed more quickly if more workers were available from other regions, but the phone grid in western New England also required extensive repairs.

THE RESILIENCE OF RESIDENTS

Residents of the affected areas coped with what the *Altoona Tribune* called "the greatest community emergency in their [Altoona and Blair County's] history" as best they could. Although crowded with people unable to go home, hospitals functioned reasonably well. In both Clearfield and Altoona, multiple babies were successfully delivered by candlelight. Altoona's hospitals were able to operate their heating equipment manually—that is, without electric power—and received intermittent power from the PRR for lights and other uses.

More generally, since natural gas service was unaffected by the storm, many residents with gas heat were able to stay in their homes. The Census Bureau listed that approximately 24 percent, or around one in four, of Altoona homes were heated with natural gas.[6] Heating units could be lit manually, and the gas company even dispatched employees to customers' homes to demonstrate the process, as detailed by the *Altoona Tribune*. The vast majority of Altoona homes, or 73 percent overall, were heated with coal. No problems with maintaining heat were reported, nor were coal shortages or delivery problems.

A few businesses used generators for electrical needs, but most used candles for lighting. As limited power returned, they supplemented candles with a few electric lights. (Although Penelec asked them to use "one bulb only," it was unclear how many businesses heeded this request.) Darkness forced businesses to curtail their hours early in the week, but they gradually returned to normal hours as more electricity became available.

NECESSITY IS THE MOTHER OF INVENTION

Electric lights, motors, and many other devices were inoperable, but residents improvised in remarkable ways, as chronicled in both Altoona newspapers. A run on gasoline and kerosene led to long lines at area gas stations. With no power, attendants dispensed fuel directly from trucks and used whatever motors they could find—bicycle motors, washing machine motors, even lawn mowers—to operate their pumps. In Clearfield, the *Progress* outlined how dairy farmers used tractors, lawn mowers, and even car motors rigged up with long belts to power their milking equipment. Back in Altoona, the Santella funeral home relit its old gas lamps, which had not been used in decades, to provide light.

Life for residents was far from normal, however. Mass transit was hobbled by local conditions and systemic disruptions from heavy snow in western Pennsylvania and flooding in the eastern part of the state. Blair County Airport closed for nearly a week, according to the *Altoona Mirror*. Almost all church services and meetings the Sunday after the storm were canceled. Courts in Clearfield and Blair Counties closed due to road conditions and lack of power. Factories closed. Even the Pennsylvania Railroad, despite its generating station, could not keep the shops open owing to insufficient power and heavy absenteeism, with more than 30 percent of workers absent.

All schools were closed. Schools in less hard hit areas managed to reopen late in the week, but Altoona schools (for power reasons) and most rural districts (because of difficulties in cleaning up trees) remained closed for the entire week. (Ironically, on Monday, the *Mirror* reported that Altoona schools hoped to open by Wednesday. That plan was quickly scuttled by Tuesday.) Unfortunately for parents in Altoona, city police banned sledding out of a concern for safety. In reality, though, the ban was unpopular and widely flaunted. In one humorous example, the *Mirror* recounted that a resident of Walton Avenue requested that the Recreation Board clear his street of sledders so traffic could move. No one in the neighborhood appreciated his effort, including drivers, as the report noted that "everyone got mad at him." The *Mirror* also described how two drunks were arrested and fined for shoving two teenage sledders near Fourth Avenue and First Street. The teenagers themselves did not get in trouble for violating the ban.

More annoying than a sledding ban for kids, perhaps, was that each day that school was closed had to be made up. Districts canceled and shortened Christmas break and other scheduled breaks in 1951. On December 13, the *Altoona Tribune* reported that Logan Township Schools, just outside of Altoona, announced plans to shorten Christmas break by one day and add days in February, March, and June to make up the lost days. Altoona City Schools, which lost five days of instruction, requested a waiver from the state, but the Pennsylvania superintendent of public instruction denied the request during the Christmas break period. Thus, on January 2, 1951, Superintendent Denniston announced a plan to cancel scheduled days off on February 22, February 28, March 21, March 22, and May 30 so the days could be made up. The school board mostly agreed but changed the last makeup day to Monday, March 26, since May 30 was Memorial Day, according to an account in the *Tribune*. (Before 1971, Memorial Day was always observed on May 30.)

Despite the lack of power, crime was not a major problem. No looting was reported, though there was shoplifting, with portable radios and batteries being the most popular items purloined. (It's unclear how widespread the shoplifting was as the *Altoona Mirror* based its report on a survey of store managers and claimed that "quite a few shoplifters were arrested," but no numbers were provided.) While there were complaints about profiteering, no charges were filed, despite bombastic threats from the Altoona mayor and local district attorney. Hollidaysburg and Tyrone declared states of emergency, which allowed police to arrest anyone for loitering, but no loitering arrests happened. Perhaps the most serious attempts at crime came in northern Altoona. Here the *Tribune* detailed two separate instances where women reported that men claiming to be Penelec representatives attempted to gain entry to their homes for "inspection" purposes. Both women denied entry to the bogus workers.

As the situation eased, businesses and residents began tabulating the costs of the disaster. Retailers, seeking to make up for lost sales during the busy Christmas shopping season, extended their hours, to 8:30 p.m., on Thursday and Friday, December 14 and 15, as announced in the *Tribune* on December 7. Frustratingly, homeowner's insurance did not cover damage from water seepage or fallen tree limbs. If high winds or hail had occurred, policyholders could file claims, but the National Board of Fire Underwriters found no evidence of either hazard. However, those with comprehensive automobile insurance coverage were able to file car insurance claims for damage from fallen trees and limbs. Despite the heavy ice and wet snow, only two roofs, both garage roofs, collapsed. The collapses occurred in Clearfield County, where more snow fell during the storm and additional snow fell early the following week.

LONG-TERM EFFECTS

Despite great disruption in the short term, long-term effects were mostly financial. Utility companies suffered great economic losses. In early December, Penelec told the *Altoona Tribune* that its loss was at least $1 million (in 1950 dollars; does not include revenue lost), and Bell Telephone provided a similar estimate. While these amounts sound high—they are equivalent to around $10 million in modern dollars—they are not enough to matter to a large utility company. Penelec simply absorbed the loss, according to a January 1951 report in the *Tribune*. The company briefly mentioned the storm as one of several factors that was expected to reduce 1950 net income by about 8 percent compared to 1949. The relatively small area affected (two counties) probably prevented more serious consequences, especially for a company planning to spend $17 million (1950 dollars) on construction projects system-wide in 1951.

Governments also had great losses, but unlike the utility companies, they could not rely on "customers" in other areas to absorb the costs. Some used money from the street or park department budgets. Others raised taxes.

In Pennsylvania numerous nuisance taxes were implemented or increased as a result of Superstorm 1950. Examples of nuisance taxes are occupation and per capita taxes, where those affected pay a fixed, often nominal, amount. Osceola Mills, in Clearfield County, and Phillipsburg, just east in western Centre County, were two municipalities that increased such taxes to cover their expenses, as chronicled in the *Clearfield Progress*. These and other post-storm tax increases persisted long after the storm.

Clearfield Borough, after a vigorous and heated debate, implemented a new $5 per capita tax to raise $17,000 for storm cleanup. Initially, Clearfield Borough Council announced a plan to consider a per capita tax of up to $10. The newspaper was vigorously opposed and wrote a strongly worded, but thoughtful, editorial arguing for a smaller tax over a longer period of years. The following day, December 10, more than 250 residents met to oppose the tax. Approximately two weeks later, on December 26, after a contentious 2.5-hour meeting the council adopted a $5 tax.

On December 3, 1951, the council voted to continue the tax for 1952. (State law at the time required the tax to be reauthorized each year.) A search of newspaper archives showed that the tax continued throughout the 1950s, with the amounts generally being $5 or $6 per year. Eventually the law was changed to allow the tax to remain in place indefinitely, so it's unclear how long the tax persisted. As of 2018, Clearfield Borough no longer had a per capita tax, but they are still common in other boroughs, townships, and school districts in the region. The growth of per capita taxes was the longest effect of the ice storm, continuing even after the forests recovered.

THE ICE ASSOCIATED WITH SUPERSTORM 1950 HAD INCREDIBLE IMPACTS IN west central Pennsylvania, primarily in Clearfield and Blair Counties. Thousands lost power for an extended period; schools, businesses, and factories closed temporarily; and daily life was greatly disrupted. Despite all of this, the general public, at least openly, dealt with the situation without complaint. There were no negative editorials haranguing municipal officials for being ill-prepared or letters complaining about Penelec. Instead, there were numerous positive editorials and advertisements praising the "community spirit" and those who helped restore services. If anything, the disaster united the community and increased local pride. The County National Bank at Clearfield, in an open letter to the readers of the *Progress*, put it best in saying, "We conclude that Clearfield is a pretty good place to live in, storm or no storm."

6

WATER EVERYWHERE

"The villain that ravaged Lock Haven has sunk back behind the mud-spewn banks that form its lair. . . . No past flood on record has sent so much water into Lock Haven so fast."
—*LOCK HAVEN (PA.) EXPRESS*, NOVEMBER 29, 1950

SUPERSTORM 1950 WAS A MULTIPLE-HAZARD EVENT THAT BROUGHT DEEP snowfall, catastrophic ice, ravaging floodwaters, damaging wind gusts, and record-setting cold. Most of the twenty-six states affected experienced only one or two of these hazards, but Pennsylvania experienced all of them. Record cold, snow, and ice devastated the western part of the state, while the eastern portion was battered by damaging wind gusts and floodwaters.

Floods kill more people and destroy more property than any other disaster in the United States.[1] As a large event with multiple hazards, Superstorm 1950 caused multiple types of floods. Inland flooding was most severe in Pennsylvania, but it also affected portions of western and northern New England. Inland flooding consisted of both flash flooding, which is a rapid flood of low-lying areas during and immediately after excessive rainfall, and riverine flooding, where streams and rivers spilled over their banks. In some areas, like central Pennsylvania, these types of floods occurred more or less simultaneously. Coastal flooding occurred along the east coast of the United States, from Maine to Delaware, and also in Canada near the western tip of Lake Erie. Strong winds associated with the storm caused a storm surge that inundated the coastline. Storm surge occurs when winds pile up seawater near a coastline, causing the ocean (or lake) to rise above normal levels. Oftentimes, larger than usual waves on top of the surge wreak additional havoc. With Superstorm 1950, damage was most severe in New Jersey and New York. In these states dozens died, and property damage was comparable to recent hurricanes.

While residents of western Pennsylvania and nearby states dug out from record-setting snow and ice, residents of the eastern part of the state and New Jersey pumped out basements and cleaned up flood debris.

PART 1

INLAND FLOODING

Inland flooding occurred in three different areas from three different causes. In north central Pennsylvania, major floods occurred along the West Branch of the Susquehanna River, both during the storm and soon after. In New England, major flooding affected western Massachusetts during the height of the storm, while additional flooding in northern New England, especially Maine, occurred during its aftermath. The orientation map in figure 8 shows major cities, rivers, and select smaller cities affected by flooding in this region. Of all the inland floods, those in Pennsylvania were the most widespread and impacted the greatest number of people.

Flash floods in Pennsylvania resulted from excessive rain. All reporting stations in the state received at least 2 inches of rain (or equivalent amounts of snow and ice), with many places receiving 3 to 5 inches with isolated higher amounts (see table 2 on p. 31). In central

FIGURE 8. Locator map of major cities, rivers, and select smaller cities referenced.

and eastern Pennsylvania, where this precipitation was exclusively rain, intense flash floods resulted. While significant property damage occurred, remarkably only one drowning death was recorded, a young hunter whose car became stuck in rising floodwaters near Duncannon, about 10 miles north of Harrisburg.

In the north central part of the state, heavy rainfall (mixed with ice and snow toward the west) caused rapid rises in small mountain streams. In Bradford, Pa., Tuna Creek and other tributaries of the Allegheny River covered several streets starting on Saturday morning, November 25; the mixed precipitation also caused a roof to collapse in nearby Kane. The *Bradford Era* provided a detailed timetable that chronicled how water rose during daylight hours, crested in the early evening, and started falling after sunset. The rise was slow enough that most businesses were able to move their stock to higher shelves or floors, so the effects were not that severe. Perhaps Bradford was spared by recent flood control measures implemented following similar floods in both 1947 and 1948.

Flooding was much worse in nearby small towns. Fifteen miles east of Bradford, Eldred was cut off from the outside world for two days due to the swollen Allegheny River, and nearby Portville, N.Y., was still flooded several days after the rain ended. Floods were also more severe about 45 miles south of Bradford, where the Elk River overflowed in St. Mary's, a city of about ten thousand. Here police and firemen plucked more than thirty people from their homes in what was considered the city's worst flood on record, as recounted in the *Philadelphia Inquirer*. Interestingly, while a similar amount of precipitation fell in Clearfield, one county to the south, the Susquehanna River did not come close to exceeding flood stage. Most of the precipitation in Clearfield fell as snow and ice, instead causing a widespread loss of electrical power.

As smaller mountain streams merged into larger rivers, the effects of flooding grew. While Altoona, Pa., was experiencing its worst ice storm on record, just 13 miles northeast in Tyrone heavy rain caused more than 5 feet of water to rush through the downtown business district, causing an estimated $750,000 to $1 million in damage (in 1950 dollars; equivalent to approximately $7.5–$10 million today). Tyrone officials used a fleet of twenty boats to evacuate more than fifty families to the high school, where more than a thousand residents and stranded travelers, including the Penn State University Marching Band (stuck on the way home from Pittsburgh), were housed during the weekend, according to the *Altoona Tribune*. To prevent looting, officials imposed martial law and summoned the National Guard. Despite the flooding and a light accumulation of ice on power lines, the electric and phone systems continued to operate as normal. Residents of Lewistown, about 35 miles east of Tyrone, were less fortunate. More than two thousand people were evacuated in what the *Altoona Tribune* characterized as the worst flooding event since 1936. Additionally, floodwaters inundated the primary electrical substation, plunging the community into darkness. Frustratingly for local citizens, this was not the first time the substation had been flooded, leading to a scathing editorial in the local newspaper and a citizen petition. The Lewistown power outage also siphoned off valuable manpower from restoration efforts in Altoona.

The worst destruction in central Pennsylvania was along the West Branch of the Susquehanna River. Renovo, a small town built on a narrow floodplain in a deep mountain valley, was (and still is) highly vulnerable to the rising river. When the river rose above flood stage on Saturday night, eighteen hundred residents, or about half of the town's population, were made temporarily homeless, according to a report in the *Philadelphia Inquirer*. Twenty-five miles downstream in Lock Haven, where the Bald Eagle Creek of Centre County merges into the Susquehanna, the situation was even worse. As floodwaters from Renovo converged with the swollen Bald Eagle Creek, the Susquehanna River rose 14 feet in just 12 hours, the fastest rise on record, reaching flood stage at approximately 7:00 p.m. Then it rose almost 7 feet more before cresting around 6:00 a.m. Sunday, as detailed in the *Lock Haven Express*. While high, the Susquehanna River remained 5 feet below the 1936 record. In contrast, Bald Eagle Creek crested at 12 feet, only 30 inches below the record set in 1936. Although no new records were established, flooding from the two rivers inundated more than 78 percent of the town with at least 3 feet of water. More than two thousand homes were flooded, forcing the evacuation of thousands of residents and sending hundreds to shelters. Contemporary reports disagreed about the exact number of people displaced, probably because accurate information was difficult to obtain. The *Lock Haven Express* itself was unable to publish for several days from the flooding.

The State Teachers College in Lock Haven (now known as Lock Haven University of Pennsylvania) was closed for the entire week; this was the longest university closure caused by the storm, other than those in snowbound areas. Its facilities were utilized for emergency services to aid the community. As the week after the storm continued, desperate city leaders requested additional water pumps to help empty the many cellars still filled with water. The city of Lock Haven and surrounding area were also unusual in that they experienced multiple severe hazards from the storm. Prior to the flooding, winds gusting as high as 77 mph overturned airplanes, unroofed buildings, disrupted power, and wreaked other havoc.

The wall of water caused additional destruction as it rolled past Lock Haven and flooded small towns downstream in Clinton and Lycoming Counties. However, the relatively large city of Williamsport (population forty-five thousand in 1950), the county seat of Lycoming County and located about 25 miles east of Lock Haven, was spared thanks to a levee. Finished just the previous summer, the levee protected downtown and much of the city from disaster, while unprotected western neighborhoods, like Newberry, and the suburb of South Williamsport across the river were severely flooded, according to a *Williamsport Sun-Gazette*. This was quite a contrast from the last major flood, in 1946, when the "entire downtown business district" of Williamsport was underwater.

The floodwaters continued downriver, forcing hundreds to evacuate in places like Milton and Lewisburg, but the effects gradually became less severe. River-level reports and descriptions in the Harrisburg newspapers indicated that at each subsequent city, the river was closer to, or below, local flood stage. Harrisburg, about 70 miles south of Williamsport, is regularly flooded by the river. Prior to this storm, moderate flooding had most recently occurred in 1946, and major record-setting flooding had occurred in 1936.[2] But Superstorm 1950 caused only minor, nuisance-type flooding in Harrisburg. There were two reasons for this. First, considerably less precipitation fell on the watershed of the North Branch of the Susquehanna River, which runs through places such as Wilkes-Barre, Pa., than on the West Branch; the two branches merge about 40 miles north of Harrisburg. For example, approximately 2.5 inches of rain fell in Wilkes-Barre, and about 1.75 inches fell in Binghamton, N.Y. Second, lower portions of the river were better able to contain the streamflow because the river channel is much wider.

Scattered inland floods occurred elsewhere in eastern Pennsylvania and New York. The effects of these were generally minor. Such floods were often associated with specific circumstances, such as a dam break near Pine Hill, N.Y., that flooded 25 square miles and washed away eight bridges, or a new dam on the Chester Creek near Philadelphia that left six hundred people homeless when it backed up. Flash flooding along the Schuylkill and Delaware Rivers in Philadelphia triggered some evacuations and even caused a few rescues, but no serious injuries or deaths were reported in the *Inquirer*.

Scattered flash floods also occurred in interior New England. One of the worst floods on record inundated North Adams in northwestern Massachusetts, where the Hoosic River jumped its banks. In nearby Lee, where 2.8 inches of rain fell, a dozen families on the north side of the town were evacuated when the Housatonic River and its tributaries overflowed, according to the *Berkshire Eagle*.

In Keene, N.H., in the southwestern part of the state, 3.3 inches of rain flooded hundreds of cellars and prompted the local milkman to use a rowboat for deliveries, as described by the *Manchester (N.H.) Morning Union*. Forty miles north of Keene, in Claremont, N.H., John McLane Clark, a local newspaper publisher, died when his canoe capsized. The calamity took place while he was surveying damage with three of his children. Fortunately his children were able to cling to low-hanging tree branches in the fast-flowing current until rescued nearly 30 minutes later. Additional flooding (and wind damage) temporarily isolated several small towns in the eastern part of New Hampshire (such as Conway and Center Ossipee). However, these disruptions were limited in scope and damage, and no deaths occurred in connection with them.

Finally, rain from the last gasps of the storm early in the following week caused flash flooding in northern New Hampshire and Maine. Basements flooded, and water ran through the lower levels of several mills in Maine, as chronicled in the *Bangor News*. The flooding was worst in far eastern Maine, where more than 5.5 inches of rain fell in Woodland (Washington County), leading to the drowning deaths of two hunters in the swollen Machias River, as described by the *Portland (Maine) Press Herald*. The heavy rains in Maine fell on saturated ground, as several inches of rain had fallen from a storm the previous week (November 20–22). Woodland itself received a whopping 12.61 inches of rain in the month of November. Overall, November 1950 was quite wet in Maine, with most cities receiving between 6 and 12 inches of rain, or roughly two to three times the normal precipitation for the month.

PART 2

COASTAL FLOODING

The interior flooding that affected the Northeast was more of a nuisance than a serious problem. In Pennsylvania, where flooding was the most widespread, there were just a few deaths and injuries from floods. Most of the costs were in property damage. Residents in those areas had weathered major floods in 1936 and 1946 and minor floods in other recent years, and they were able to minimize the risks. In contrast, the coastal flooding along the northern Eastern Seaboard caused widespread destruction and dozens of deaths. Destruction occurred from Maryland to Maine as strong easterly winds created 20- to 40-foot waves that smashed the coastline. New Jersey suffered the most; every coastal town in that state suffered damage comparable to the Great Atlantic Hurricane of 1944. More notably, dozens of people drowned from coastal floods, making New Jersey one of the states with the most storm-related deaths.

Like elsewhere, the storm was poorly forecast and not expected to affect the area. However, in New Jersey the poor forecast directly caused deaths, as people in beach cottages and hunters in marshlands were completely unprepared. In a widely covered story, the Clarence Owens family spent Friday night visiting his father at his riverside cottage along the Maurice River in Cumberland County. The next morning, the family awoke to the cottage being washed away—with them inside. The fast-moving water was so strong that the adults were barely able to save themselves. Four children and an adult of unknown relation drowned. Elsewhere in southern New Jersey, dozens of hunters and others went missing; many were listed in the *Vineland (N.J.) Times-Journal*. Most were never found.

Quickly rising water forced numerous other evacuations and rescues. Near the mouth of the Delaware River, three hundred families were evacuated from Penns Grove and Pennsville, N.J., both across the river from Wilmington, Del., and at the north end of Delaware Bay, as recorded in the *Wilmington (Del.) Journal-Every Evening*. Additional evacuations occurred in hard-hit Cumberland County, in southwestern New Jersey, where more than twenty-five hundred residents were displaced and nearly a thousand ended up in shelters, according to the *Vineland Times-Journal*. Surprised hunters were at greatest risk; the *Philadelphia Inquirer* described how nearly a hundred were rescued in southern New Jersey, where the Coast Guard used helicopters to help spot parked automobiles near marshes. Further north, at Island Beach, N.J. (located near Toms River), the Coast Guard used a duck boat to rescue a 44-year-old father and his 8-year-old son trapped by rising water. They had arrived late Friday night to go duck hunting on Saturday morning but awoke to find themselves surrounded by rising floodwaters, as portrayed by the *Asbury Park Press*.

Coastal destruction also occurred in Maryland and Delaware. Damage was heightened because the storm occurred near the time of spring tide—Friday, November 24, featured a full moon—and this increased the storm surge impacts.[3] In Maryland, flooding occurred along all coastlines. Floodwaters from Chesapeake Bay cut off Miller's Island, just outside of Baltimore, and across the bay the oyster industry in Queen Anne's County suffered more than $100,000 in damage to boats and shacks, as reported by the *Wilmington Journal Every-Evening*. Flooding occurred along the Delaware coastline as well, and further south on the Delmarva Peninsula, Ocean City, Md., was inundated. In New Jersey, the *Vineland Times-Journal* described how most homes in Moores Beach and Thompsons Beach (two now-abandoned communities southeast of Port Norris) were destroyed by the water. More than 120 homes, many of them vacation cottages, simply washed away. Some larger structures, though, floated to new locations. For example, in a story from the *Philadelphia Inquirer*, Mrs. Ursula Johnston, a Philadelphia resident, went to check on her two-story cottage in Thompsons Beach after the storm. Much to her dismay, she discovered that her cottage was gone. She started back to Philadelphia, only to spot her cottage—standing upright and apparently undamaged—on marshland located 4 miles from its foundation. Damage to the structure and furnishings was minimal. She planned to call a contractor to return it to its original location.

Destruction in small towns along the New Jersey side of Delaware Bay, especially near the Maurice River, was severe. Most affected were the impoverished residents of Bivalve and Shell Pile, oyster industry company towns with large African American populations.[4] Forced to evacuate as water flooded the company-owned houses, many of the residents had to abandon all personal belongings. Luckily, most of the buildings survived, and no worker deaths were recorded. Interior damage was primarily on the first floors, and workers generally lived on second floors, according to the *Vineland Times-Journal*. Shucking and packing operations resumed as early as Tuesday, as reported by the *Millville Daily*. In another fortunate break, the storm also did not wipe out the oyster industry, as initially feared. While there was plenty of flood damage, the oyster beds were not silted over, so a $5 million bullet was dodged. Unfortunately, the arrival of the pathogen MSX a few years later would kill nearly all the oysters, devastate the industry, and turn the communities into ghost towns.

As mentioned, the worst damage overall was in coastal New Jersey. Strong easterly winds, which were perpendicular to the mostly north–south oriented shoreline, caused the highest tides in this region, and the gusty winds themselves caused additional damage. Every coastal town experienced wind and water damage from the storm, and in many places this damage was comparable to that of the Great Atlantic Hurricane just six years earlier. Boatyards were destroyed, boardwalks were damaged, roads washed away, and new inlets were cut. The *Philadelphia Inquirer, Asbury Park Press,* and *New York Times* editions immediately following the storm were filled with individual reports from specific communities.

A full accounting of the coastal damage could easily fill an entire chapter; only the most notable stories follow.

Two-thirds of Ocean City, N.J., was flooded by the storm, and electric power was lost throughout the town. In Atlantic City, thousands of tourists were marooned when all major roads and railroad lines were washed out. It took days to reestablish rail service, and overall damage to the town was at least $500,000, according to a *New York Times* report. Farther north, a large excursion boat, *The City of New York*, broke free from its moorings and was carried onto a front lawn in Keyport. From a close examination of the AP wirephoto, it appears that the boat was at least 30 feet tall and 100 feet long. In nearby Morgan Beach, all but two of thirty-eight homes were destroyed, creating two hundred homeless residents, while in Lawrence Harbor fourteen homes were uplifted from their foundations and moved inland. Finally, in Long Branch, N.J., tragedy struck when a 200-foot section of boardwalk disintegrated under a barrage of wind-driven 25-foot waves. Hurled by the wind to a height of more than 20 feet into the air, debris rained down on a crowd that had gathered to watch the storm, killing two and seriously injuring at least half a dozen more.

Further north, significant coastal flooding and disruption to shipping occurred in and around New York City, where the *New York Times* described how every available Coast Guard ship was used to help stabilize barges and small craft torn from their moorings. Hazardous docking conditions forced the suspension of municipal ferry service to

Staten Island for the first time in history, and ferry service to New Jersey and to Connecticut (across Long Island Sound) was also halted. (Most service resumed on Sunday and was fully operational by Monday.) In Queens, a dike restraining Flushing Bay failed, flooding LaGuardia Airport with 3 to 12 feet of seawater. The airport was closed for the better part of a day as a result. (It may have closed anyway; most airports in the region closed due to high winds.) Extensive coastal flooding occurred along the eastern shore of Staten Island. The worst occurrence was along the southeastern coast, from Fort Wadsworth to Princess Bay, a distance of about 15 miles. Here, 2 to 6 feet of seawater swept as far as a half mile inland. The *New York Times* specified that fifteen hundred families were evacuated.

Coastal damage and flooding also tormented residents of Connecticut. Conditions were worst in western Connecticut, where the coastline arcs to the southwest and thus has more exposure to winds coming from the east. Here, persistent easterly winds piled excess water in the western portion of Long Island Sound. Greenwich and Stamford, the two most southwestward towns in the state, suffered the most destruction. The *Hartford Courant* reported that high water caused more than seventy-five Greenwich families to flee their homes, and the town's bathhouses at Island Beach were destroyed. In Stamford, floodwaters were 5 feet deep and inundated the first floors of most residences on Shippan Point, a peninsula that extends into Long Island Sound. However, coastal damage in the rest of Connecticut and Rhode Island, where the coastline faces more southward and less eastward, was generally light. Despite the high waves, contemporary accounts from Rhode Island and eastern Massachusetts noted how *little* damage occurred along the coastlines of Connecticut, Rhode Island, and Massachusetts. The only major damage in this region was at Saybrook Dock, which was destroyed by high waves. However, it was more vulnerable than most other structures because of its location on the western shore of the Connecticut River's estuary. The *Hartford Courant* noted that a traffic jam occurred the following day as hundreds of people drove out to see the remnant of the dock.

Damage along the New Hampshire and Maine coasts was more severe than in southeastern New England since these coasts mostly face east or southeast. Large waves generated by hurricane-force winds offshore wreaked havoc with ships. Outside of Portsmouth, N.H., a 4,100-ton Norwegian freighter was driven aground, as described by the *Manchester (N.H.) Morning Union*. The *Boston Globe* told how a Coast Guard cutter was dispatched when rough seas caused cargo to break loose on the SS *Ames Victory*, threatening to capsize the ship. The ship was bound for New York and was located 500 miles east of Ambrose Lightship (New York's harbor) when the mayday call went out.

A HEROIC RESCUE

A dramatic rescue occurred off the coast of Nantucket, an island about 25 miles south of Cape Cod in eastern Massachusetts. The U.S. Navy destroyer escort 532, under the command of Lt. Calvin Runnells, was being towed to Maine for recommissioning when it broke free of the tow rope near Nantucket. Twelve men now found themselves marooned on a ship with no working mechanical systems. Worse yet, 35- to 50-foot seas and 70-knot winds battered the ship as if it were a child's bathtub toy. After being alerted to the precarious situation, the *Nipmuc*, a tugboat from Newport, R.I., was dispatched to rescue the men and save the ship, if possible. After fighting the rough seas for 13 hours, the captain of the *Nipmuc*, Lt. William J. Bryan, located the stranded ship. Amazingly, he managed to connect a tow cable despite the ferocious waves and wind. In an account in the *Portland (Maine) Press Herald*, Lt. Runnells praised Lt. Bryan, saying he had "never seen a more brilliant job of seamanship in his life. . . . [It was] as though we were in a windless bay." He credited Lt. Bryan with saving the men's lives.

Damage was substantial along the rocky coastline of Maine. Grandstands at both Old Orchard Beach and Cumberland were destroyed by wind and waves, and the *Bangor News* also reported that the Ogunquit Playhouse suffered major damage. Just offshore, more than fifty buoys along the northern New England coast were broken, moved, or simply vanished, according to the *Boston Globe*. Lighthouses were also battered. Worst damaged was the Matinicus Rock lighthouse, about 25 miles south of Rockland, Maine. Seawater flowed into the building, ruining all machinery and knocking the lighthouse out of commission. The *Portland (Maine) Press Herald* also noted that waves smashed the glass panes at the top of the tower, an astounding 107 feet above sea level. Sea water also seeped into freshwater storage tanks, forcing the three keepers to survive off distilled water meant for batteries. The men were finally rescued four days after the storm. Persistent heavy seas in the storm's wake precluded the Coast Guard from reaching the stranded men sooner.

Unlike the oyster industry in New Jersey, the lobster industry in Maine suffered catastrophic losses. Lobster traps are usually unaffected by storms, but the ferocity of this storm caused lobstermen to lose 70 to 100 percent of their traps (as well as the contents of such traps), as detailed in the *Portland (Maine) Press Herald*. Boats and other lobster equipment were also heavily damaged. The direct loss of traps and equipment was estimated at $1.35 million in 1950 dollars (or around $13 million in modern dollars). Industry spokespeople quoted in the *Press Herald* based that estimate on 175,000 traps lost (at $7 each), plus another $100,000 in damage to boats. The loss prompted a rare federal response. The Reconstruction Finance Corporation, a New Deal agency established during the Great Depression, set up a local loan office to help finance replacements. However, Maine's governor felt that the lobstermen needed urgent relief and asked the Red Cross for immediate help.

Finally, major coastal flooding also occurred at an unexpected location hundreds of miles inland from the storm: the southwest corner of Lake Ontario. Near Hamilton, Ontario, easterly winds caused waves that the *Ottawa Journal* estimated were 30 to 40 feet; the *Buffalo Evening News* reported that the largest waves were nearly 60 feet. Whatever the actual size, these giant waves battered beachfront communities in the area, sweeping twelve homes away entirely and damaging many others. Thankfully no deaths were recorded, but at least 125 people required medical treatment for injuries. More than 300 were made homeless, based on the *Ottawa Journal*, or 500 people based on the *Buffalo Evening News's* report.

THE INLAND FLOODING ASSOCIATED WITH SUPERSTORM 1950 WAS RELATIVELY small in terms of the size of the area affected, but it was major where it occurred. Still, the excessive rain did help to replenish low water supplies. (Eastern New York and New England were still experiencing the effects of a drought dating back to 1949.) According to the *New York Times*, the reservoirs that feed New York City received an approximate twenty-five-day supply of water from the two days of rain; capacity after the storm was roughly double what it was a year prior. In an ironic story, a work crew was seeding clouds in the Catskills on the Saturday of the storm to create some much needed rain. At 5:00 p.m. they were ordered to cease (and told to not resume on Sunday), not because of all the rain falling but because they were needed to help shore up area dams due to fast-rising waters.

The coastal flooding caused record tides along the New York and New Jersey coastlines, along with numerous deaths. The storm was widely compared to the Great Atlantic Hurricane of 1944 in New Jersey and the 1938 Great New England Hurricane elsewhere. Both of these storms, like all landfalling hurricanes, caused extensive wind damage near and far from the coast. Superstorm 1950 also caused record, hurricane-force winds and extensive wind damage near and far. The next chapter describes the windstorm and its effects.

This iconic image shows snowbound Webster Avenue against the Pittsburgh skyline. (AP photo/Walter Stein.)

Snow-covered cars on Main Street in Weirton, W.Va. (Courtesy Weirton Area Museum and Cultural Center.)

In Weirton and most other snowbound areas, walking was the only way to get around. (Courtesy Weirton Area Museum and Cultural Center.)

People walked great distances through the snow. These men walked 4 miles through hilly Pittsburgh to buy groceries, then hauled them another 4 miles home. (Courtesy *Pittsburgh Post-Gazette*.)

Most employees at Weirton Steel walked or rode the bus to work. The *Weirton Steel Bulletin* proudly proclaimed that "not a ton of much-needed steel production was lost." (Courtesy Weirton Area Museum and Cultural Center.)

Snow and icicles covered the American Bridge Company in Ambridge, Pa. (Photo by Butch O'Keefe.)

Pittsburgh dumped excess snow in the Monongahela River, a common practice for cities in 1950. (Courtesy *Pittsburgh Post-Gazette.*)

To hasten snow cleanup, the city of Pittsburgh prohibited nonessential vehicles from downtown and the National Guard staffed forty-three checkpoints. Motorists needed an orange card or strong justification to pass. (Courtesy *Pittsburgh Post-Gazette.*)

The Ohio State–Michigan football game quickly became a farce as bitter cold, snow, and wind made it extremely difficult to advance the ball. (From The Ohio State University Photo Archives, drawer 159, folder 35.)

Near-zero visibility during the Ohio State–Michigan game made it difficult for spectators in upper sections to see the field. (From The Ohio State University Photo Archives, drawer 159, folder 35.)

Ice and snow felled this transmission tower in the Collinsville neighborhood of Altoona, Pa. (Courtesy *Altoona Mirror*.)

Two linesmen examine an ice-encrusted transmission line near Altoona. (Courtesy *Altoona Mirror*.)

Fallen limbs blocked Howard Avenue in Altoona. (Courtesy *Altoona Mirror.*)

These poles in the Lakemont area of Altoona remained standing despite the incredible weight of more than a dozen ice-covered wires. (Courtesy *Altoona Mirror.*)

Volunteers in Lock Haven, Pa., assisting in rescues during the flood. (Courtesy Annie Halenbake Ross Library, Lock Haven, Pa.)

Destroyed boardwalk in Long Branch, N.J. (Courtesy Local History Room, Long Branch [N.J.] Public Library.)

Wind damage to the University of Vermont gymnasium. (From Silver Special Collections, University of Vermont Libraries.)

7

BLOWN AWAY

"Never in the history of the insurance business in the eastern area has there been such a terrific loss."
—VERMONT STATE ASSOCIATION OF INSURANCE
AGENTS, NOVEMBER 29, 1950[1]

SUPERSTORM 1950 BROUGHT CRIPPLING SNOW TO THE OHIO VALLEY AND record cold to the Southeast. These extreme conditions were unavoidable and notable for both their ferocity and their widespread nature. It didn't matter if a person was rich or poor or lived in a large city or on a rural homestead; life was greatly disrupted.

The triad of localized disasters that affected the Northeast—ice, flooding, and wind—had more variable effects. An incredible ice storm disrupted utility service for weeks in Blair County, Pa., but residents of neighboring Huntingdon County experienced merely sporadic outages. Flood waters ravaged South Williamsport, Pa., but across the river the city of Williamsport was protected by a new levee. Wind brought down trees in many areas, especially Vermont and eastern New York, but if a tree did not happen to strike you, your house, your car, or a nearby electric line, the disruption was short-lived and generally mild.

Although hurricanes are less regular in the Northeast than in other parts of the county, two major hurricanes battered the region in the twelve years prior to the storm. The Great New England Hurricane of 1938 was the most powerful and deadliest hurricane to strike the region, killing close to seven hundred people. It remains the benchmark for comparison with other coastal storms. Six years later, the Great Atlantic Hurricane of 1944 hugged the coast from Cape Hatteras northward and eastward before passing through Rhode Island and southeastern Massachusetts. New Jersey was hardest hit by this storm.

Hurricanes in the tropics form under very different conditions than those that form midlatitude cyclones, and Superstorm 1950 was not a hurricane. Yet in many places, local residents compared its damage to recent hurricanes; the *Addison County Independent* repeatedly termed it a hurricane in its coverage (even though it noted that the Weather Bureau disagreed.). Although the death toll was lower, in most places the damage was equivalent to, and in some places worse than, the damage from the hurricanes of 1938 and 1944.

A DE FACTO HURRICANE

As previously explained, an unusually strong high pressure near Labrador prevented Superstorm 1950 from moving offshore as it developed. As the storm intensified, its pressure decreased, causing a very large difference in pressures between the storm and high pressure center, also called a "pressure gradient." Strong pressure gradients cause fast wind speeds, and this one was so intense that gale-force winds with frequent gusts to hurricane-like speeds affected a widespread area. Eastern Pennsylvania, New Jersey, New York, and every New England state experienced hurricane-force (74 mph) wind gusts or stronger. Table 3 (p. 31) lists some of the peak wind gusts from the storm.

Unsurprisingly, the effects of such strong winds were similar to those of an actual hurricane: hundreds of thousands of homes lost shingles, plate glass windows were blown out at businesses throughout the region, and falling tree limbs and entire falling trees caused power outages and disrupted phone service for millions. Falling objects injured scores of people and directly killed some unlucky others. No town or city from Chesapeake Bay to Maine was spared, though the worst destruction was concentrated in eastern New York and Vermont. Notable damage also occurred in Canada. In eastern Ontario and Quebec, eleven people died. Most of the Canadian deaths were in traffic accidents on what the *Ottawa Journal* termed "wind-glazed roads."

Allentown, Pa., in the eastern part of the state, experienced hurricane-force winds that gusted to 88 mph. Many plate glass windows exploded and billboards fell in that part of state, especially southward in the Philadelphia area, where there were many damage reports. Accounts of storm damage in New Jersey tended to focus on the coastal flooding, though there were many wind reports as well. The *Vineland Times-Journal* claimed the wind damage in the state was worse than with the 1944 hurricane.

In Harrisburg, Pa., wind split the 30-foot Christmas tree in Market Square, and only the guy wires prevented it from crashing down, as described in the *Evening News*. In Tamaqua, 75 miles northeast of Harrisburg, the wind destroyed a wooden roller coaster at Lakeside Park. Twenty miles west of Tamaqua, in Llewellyn, residents were warned to watch out for what the *Philadelphia Inquirer* termed "vicious" hungry minks after one hundred escaped from a local farm. The wind had destroyed their cages. Both towns are in Schuylkill County, which experienced the greatest number of utility problems in Pennsylvania Power and Light's twenty-eight-county service area.

In Scranton, Pa., 100 feet of newly constructed wall was blown over at the future site of Memorial Stadium, prompting a school board investigation into possible shoddy construction practices. Like the misadventures with the "destroyed" Erie snow loader, this story appears to be a case where the storm was used to settle an old score. The Democratic-led school board initially misstated that the collapse was at the old stadium. As a result, Republican director Nelson Nichols told the *Scranton Tribune* that they were covering up shoddy construction practices at the new site. He alleged that the collapsed wall had large cracks and was missing a concrete footing, and he demanded an immediate investigation. The architect, engineer, and contractor determined that the wall collapsed because of "unusually excessive wind velocity which caused tremendous damage throughout the entire area." Furthermore, the wall was built in line with the plans (though some specifications had been altered) and the "workmanship was good." The findings exonerated the board. When asked about his thoughts on the findings, director Nichols refused to apologize, claiming that he only stated "facts" based on what he observed and arguing that changes in specifications were leading to an "inferior" wall.

Wind-driven rain caused additional damage to structures, most notably the Berks County courthouse in Reading, where numerous windows in the upper floors were blown out. Wind-driven rain blown through *walls* caused so many complaints in the Philadelphia area that the local builders association bought a large advertisement in the *Inquirer* on Tuesday, November 28. The ad touted their "sound construction" and blamed problems with wind-driven rain on "one of the most severe storms in the history of Philadelphia."

The combined effects of high winds and heavy rains damaged road infrastructure in the state as well. The last covered bridge across the Delaware River, connecting Portland, Pa., with Columbia, N.J., was damaged by the storm, as mentioned by the *Asbury Park Press*. It would be repaired, but Hurricane Diane in 1955 ultimately destroyed it. In the northeastern part of the Pennsylvania, more than $250,000 in damage occurred in the Scranton area: a combination of thousands of fallen trees, blown down snow fences, and a two bridges that were washed away. The *Tribune* noted that one of the bridges, in Susquehanna County, was a 90-foot concrete span. In Harrisburg, construction on the M. Harvey Taylor Bridge, a 4,219-foot span across the Susquehanna River, was delayed when earthen roadways used to move equipment were washed away.[2]

The windstorm caused five deaths in eastern Pennsylvania and Maryland from traffic accidents. A 61-year-old man was electrocuted when he drove around a roadblock and into a live power line near Towanda, Pa., in the northern part of the state. Four other people died in more conventional traffic accidents, one near Philadelphia and three in Maryland. In Maryland, the *Baltimore Evening Sun* chronicled how temperatures fell so quickly that standing water on roadways froze to ice. Additionally, numerous car locks froze, causing people to get stuck wherever they were.

Carbon monoxide sickened several people in the Scranton area. Several chimneys became plugged with bricks and other debris blown in by wind and/or washed in by rain, causing exhaust gases to back up into structures. However, the *Tribune* reported that all of the victims were saved in time.

THE FALLEN-TREE EMPIRE IN THE EMPIRE STATE

Buffalo, N.Y., has a reputation for cold and intense lake effect snow. Much of this is the result of the famous Blizzard of 1977, but the city had this reputation in the 1950s as well. However, Superstorm 1950 spared the Buffalo area. The *Buffalo Evening News* correctly noted that damage "on every side" of Buffalo was worse. Flooding occurred to the west and south, snow to the south-southwest, and wind damage to the east and north.

Wind damage was pervasive across the eastern two-thirds of New York State. In central New York, where winds gusted to 90 mph at Syracuse, the wind felled ten to fifteen thousand trees according to the *Post-Standard*. Tens of thousands of residents were plunged into darkness and schools were closed on Monday, November 27. Power restoration was slow. As of 4:00 p.m. on Sunday, November 26, more than ten thousand residents still did not have power. Despite much of the southern part of Syracuse not having electricity on Monday night, schools reopened on Tuesday. East and north of Syracuse, millions of trees were flattened in the Adirondack mountains, in what a local historian characterized to the *Buffalo Evening News* as the "worst storm in recorded history." Many hunters were trapped in the woods, with more than a hundred in the Saranac Lake area alone. Eleven rescued hunters from the Syracuse area told the *Post-Standard* that tree damage along trails in the Adirondacks was the worst they'd ever seen.

The cities of Albany and Troy, in the eastern part of the state, suffered the worst damage and experienced the longest recovery time, as indicated by reports of the number of customers without power. In Troy, a falling tree crushed Gallo's Fruit Store, and widespread damage occurred in the city and adjacent rural areas. The *Troy Times-Record* also reported that several state government buildings in Albany were damaged. Schoolchildren throughout the region had the day off on Monday, and Troy schools remain closed until Wednesday, November 29. This was the longest district-wide school closing in the Northeast outside of Vermont; most districts did not close at all. (That said, there were instances in this region where a school building itself sustained heavy damage and had to be closed until repairs could be completed.)

Upstate New York also suffered more than fifteen deaths according to UPI syndicated reports. About one-third involved traffic crashes, and another third involved falling trees or other objects. For example, a flying concrete block killed a man near Oswego, and another man died when a flagpole struck him. The remaining deaths were from a variety of causes. Cleaning up damage was hazardous; one person suffered a heart attack, while another was blown off a roof. A live wire electrocuted a farmer in Copake, N.Y. Most of the people who died were middle-aged or older, and the majority were male.

Damage in Downstate New York, especially in New York City itself, was also notable. Winds at both major airports in New York City peaked at 94 mph. The high winds wreaked havoc throughout the city. The *New York Times* succinctly summarized the destruction in its edition the day after the storm:

> [New York City] looked as if a giant Halloween prankster had been on
> a spree in it. Store windows broke, buildings collapsed, trees toppled,

cornices and roofs tore loose from buildings, and power lines snaked crazily through some streets.

The *Times* also provided many statistics to illustrate the magnitude of the damage. More than 407,000 customers lost power because of downed wires; 419 utility poles were blown down and an estimated 2,464 electrical and telephone wires were broken. New York City police also counted 1,333 damaged signs and 3,060 broken windows.

Falling objects were a significant hazard in the vertically oriented city that featured widespread masonry construction in 1950. Police estimated that 721 building cornices were damaged, some of which turned into serious hazards. Abraham Yaeger, 27, was killed when a 1-ton piece of a building at 33 Union Square West fell on him. Another falling cornice injured five people and damaged twenty-one automobiles on West Forty-Fourth Street. Besides flying debris, the *Times* also described deaths in the region caused by heart attacks, electrocution, and falls from roofs. In total, at least 44 people, primarily men, died in the New York City metropolitan area, with 5 of those within New York City itself. Another 345 people were injured in New York City proper. (No injury statistic for the metropolitan area was available.)

Flora were also badly damaged; more than twenty-seven hundred trees in New York City were downed. The Bronx Zoo closed for the first time since 1896 as more than fifty trees were strewn about on paths and roads. Luckily, two magpies that had fled the zoo during the storm returned. Their cage had been shattered by falling trees.

VERMONT'S WORST WIND DISASTER

In western New England, wind damage was similar in magnitude to that in eastern New York State. Connecticut suffered its worst disaster since the 1938 hurricane. Winds in Hartford gusted to 100 mph, and the *Hartford Courant* said that every town in the state suffered wind damage. In Massachusetts, wind damage was most severe in the west.

The area around Pittsfield was considered among the worst-damaged in Massachusetts according to the *Berkshire Eagle*. Beckett and North Adams schools closed due to wind damage, as well as those in nearby New Lebanon, N.Y. Other area schools closed because of flooding.

Vermont experienced the worst wind damage in New England and arguably the entire Northeast. The governor called the storm the worst disaster in the state since a flood in 1927, Green Mountain Power described it to the *Burlington Free Press* as the "worst, most complete disaster ever" in the company's history, and the Central Vermont Public Service Corporation told the Associated Press that the damage was worse than in 1938. New England Telephone ranked it as the worst storm for Vermont and added that more than half of the utility's $825,000 in losses occurred in that state. Reconnecting the estimated fifteen thousand phones without service took time, too; three days after the storm, eleven towns in Vermont still had no phone service at all, according to reports in the *Burlington Free Press*. This was more than the rest of New England combined, where only six other towns lacked phone service at that point. (Sixty-seven towns had lost service at the peak of the storm.) Considering all the damage, it is remarkable that only one person was killed directly by the storm in Vermont. The *Burlington Free Press* described how Roy Bean, a section foreman for Rutland Railroad, was fatally injured when his motorized handcar derailed. The derailment happened because the wind blew a building onto the railroad tracks.

Towns in the west central part of the state, especially from Burlington south to Rutland (where four to five hundred buildings were damaged), experienced the most extensive damage. After touring the damage, the governor described it as "far worse" than that of the 1938 hurricane. In Brandon, about 15 miles north of Rutland, damage was considered the greatest ever in the history of the town, and a particularly strong gale toppled the steeple of the Baptist church, sending it crashing through the roof and sanctuary into the basement. Just north, in Addison County, more than a thousand cattle were crushed by collapsing barns or had to be shot after the storm due to injuries, according to multiple reports. Middlebury, the county seat, and its eponymous college suffered $150,000 and $100,000 in damages, respectively. Morgan School in East Monkton and Brandon School (in Rutland County) were among the schools expected to be closed for an extended period due to substantial roof damage. Elsewhere in Vermont, a furniture plant in Randolph was crippled when a key building housing its ventilation system was destroyed, while in Burlington, the roof was ripped off a 2-year-old building at the Champlain Valley Fairgrounds, damaging or destroying fifteen cars. In Lyndon, a covered bridge was blown away. Numerous other reports of damage from Vermont listed roofs ripped off buildings, blown out plate glass windows, flattened outbuildings, and uncountable trees and power lines down.

Forest damage in Vermont was so severe that area foresters struggled to quantify it. Gerald Wheeler, supervisor of Green Mountain National Forecast, believed that 20 million board feet of lumber, worth $1.5 to $2 million (or $15 to $20 million today), were destroyed. A. W. Gottlieb, deputy state forester, told the *Bennington Evening Banner* that it could be as long as a year before an accurate estimate would be made.

UVM STUDENTS STEP UP

At the University of Vermont, in Burlington, $150,000 in damage to buildings and equipment was estimated from wind alone, not including tree damage or water damage done to buildings that lost roofing or windows. An interesting story of service also originated from the university. More than two hundred students volunteered to clean up damage on farms throughout Chittenden County, and crews worked one or more days from 7:30 a.m. to dusk on the Thursday through Sunday following the storm. The groups of students were organized by a dean and supported by a food truck supplied by the university canteen. (Although not explicitly stated by the *Burlington Free Press*, presumably they were excused from classes.)

WIND DAMAGE ELSEWHERE IN NEW ENGLAND

Wind damage in eastern New England was less severe than in western New England. Peak wind gusts here were slightly lower, and Rhode Island and eastern Massachusetts escaped with only minor wind damage noted in the *Providence Journal*. In New Hampshire and Maine, trees were blown down and many people lost power or phone service, but recovery time was quicker. The death toll was higher, however. Two Maine residents died in traffic accidents following the storm. In Walpole, N.H., George Cobb, a well-known farmer, was struck by lightning while making repairs to the roof of a chicken house. His death, which was detailed in the *Manchester (N.H.) Morning Union*, occurred as part of the lingering rains that flooded parts of northern New England after the storm. (Recall from the previous chapter that two hunters in Maine drowned from these rains.) Mr. Cobb's death, on Tuesday, November 28, was the last storm-related death in New England.

There were fewer examples of major destruction in Maine and New Hampshire, but several notable things happened. In Concord, N.H., the roof blew off the A. Perley Fitch Drug Company warehouse, causing $250,000 in building damage and an additional loss of $250,000 in materials, according to the *Manchester (N.H.) Morning Union*. (Adjusting for inflation, the combined losses are around $5 million in modern dollars.) The damage to the warehouse affected hospitals statewide, as the Fitch Drug Company supplied all of them. In Pembroke, the roof blew off the "sanitorium," setting off the sprinkler system and flooding the building. Thankfully, none of the thirty-five patients was injured. In Maine, the *Portland Press Herald* reported that the screen at the Kennebunk Drive-In, considered the largest in the state, was blown down for a loss of $6,000. Also in Maine, a 150-year-old elm tree planted by Hannibal Hamlin when he was a youth was blown down. Hamlin was Abraham Lincoln's first vice president, and the only Mainer to serve in that office.

INTERLUDE

THE WINDSTORM'S EFFECTS ON SPORTS

Like in the other regions, sporting events in the Northeast were affected by the storm. Most football games were still played, however. Only the Colgate–Rutgers game (in New Brunswick, N.J.) was canceled, and this was because flooding made it difficult to access the stadium. This was the first time since Rutgers began play in 1869 that a game was canceled due to weather. The Niagara–Scranton game was postponed to Sunday, when Niagara defeated the home team 12–0 on a field termed an "ice rink" by the *Scranton Tribune*. Another game that was supposedly postponed, Kings College (of Wilkes-Barre, Pa.) at Gannon University (Erie, Pa.), was actually canceled on Tuesday, November 21, four days before the storm. The cancellation was for reasons not related to the storm, but given that Erie received more than 2 feet of snow from the storm, the game may have been canceled anyway.

Most football games were muddy, sloppy affairs. In Philadelphia, Cornell upset Penn 13–6, in what the *Inquirer* termed a "quagmire" under "intolerable" conditions that was a "tug of war" instead of football. There were more than seven turnovers, and like their football brethren playing in Columbus, Ohio, both teams resorted to punting the ball instead of trying to score (twenty-six punts occurred). While a crowd of 17,846 fans sounds impressive, this was only one-third of those who purchased tickets. The *Inquirer* editorial board cautioned that if colleges continued to play under such conditions, people might watch games on television instead of coming to stadiums. In western New Jersey, Princeton defeated Dartmouth 13–7 before five thousand hardy fans instead of the thirty thousand expected. A 60 mph wind-driven rain soaked everything and made the ball slippery, resulting in almost as many turnovers (eight) as first downs (nine). At the massive Polo Grounds in New York City, so few people watched Fordham shut out New York University (final score 13–0) that a *Boston Globe* reporter observed that all of the fans could have fit in one section "with alternate seats vacant for comfort."

Two games still drew large crowds despite miserable conditions. One was the Canadian Football League's championship game for the Grey Cup. Held in Toronto, a record 27,100 fans ventured to Varsity Stadium. After watching the game, some of those fans probably regretted their attendance. Eight inches of snow had fallen on Friday, but warming temperatures on Saturday caused the snow to melt and saturate the field. Additionally, a cold rain fell through the entire game to make everything wetter. The field became a "sea of mud and water," and was made more hazardous from deep ruts left behind by snow removal machines. Ironically, these machines were used since game organizers didn't feel that the potential for bad weather was sufficient to justify spending $8,000 (in 1950 Canadian dollars) on a tarp, according to the *Ottawa Journal*. (Though the falling rain would have made the field muddy anyway.) Despite completing just three forward passes, Toronto prevailed, 13-0, over Winnipeg. All scoring came off Winnipeg turnovers. Farther east, the other game with a large crowd was played in much drier but windier Cambridge, Mass. Here, in the annual playing of "The Game," Yale defeated rival Harvard 14-6 in another fumble fest. But the wind didn't deter 40,000 spectators from attending—it was one of the largest crowds of any game that day.

SOCIETAL IMPACTS OF THE WIND

In additional to property destruction, the windstorm had many other effects. Most were similar to those experienced in snow- and ice-bound regions. Some schools closed, but the closings were shorter. Most closed schools were in the hardest hit areas (eastern New York and Vermont) or were isolated cases where the school building itself was badly damaged. Colleges all stayed open, though they were more lenient with attendance policies. The narrative was the same for industry and transportation; unless specifically damaged by the storm, factories remained open and transportation disruptions were scattered.

Residents and businesses dealt with power outages in a manner similar to that of folks in the ice region. People used whatever gasoline motors they could rig to power gas pumps and other machines, with reports found in multiple locations. Stores detailed increased sales of gas lanterns, flashlights, and candles. (In Burlington, Vt., birthday candles were particularly hot sellers.)

Medical facilities continued as best they could without power. Babies were still born, but by candlelight. In Plymouth, N.H., a healthy baby boy was born in the local hospital by the light of stable lanterns and battery-powered lamps, according to the *Boston Globe*. Multiple reports described how three babies in Burlington, Vt., were born under a combination of candles and battery lighting. In Nashua, N.H., firemen used manual power to keep a sick 13-year-old boy alive until a generator could be set up to power his iron lung. Such stories were compelling but much fewer in number compared to those in snowy areas.

The two most unusual effects of the storm were those on crime and insurance. Unlike snowbound locales, where crime rates dropped, there was little reporting on any relative changes in crime. Nonetheless, police and firemen spent a busy weekend providing aid, directing traffic, and dealing with the chaos resulting from the storm. The *Philadelphia Inquirer* called Saturday November 25 the "busiest day on record" for police; the *New York Times* had a similar report. In Boston, police stationed themselves at more than fifty stores with blown out windows to prevent looting. Although the *Globe* did not report any looting there, it did report on a theft in Newburyport, Mass., about 35 miles northeast of Boston. A thief there used a smashed plate glass window to help himself to seventeen watches worth $305.

A more serious crime of opportunity happened in nearby Marblehead, Mass., about 15 miles northeast of Boston. Beryl Atherton, a 47-year-old single schoolteacher, was murdered. The killer slashed her throat six to ten times. Police believe she was preparing dinner on

Saturday evening when it happened. The storm had cut power, and the wind was howling loud enough to cover up any screams, according to the *Manchester (N.H.) Morning Union.* The scene was not discovered until Monday morning by the milkman. Police theorized to the *Boston Globe* that the killer came back and cleaned up, again taking advantage of the blackout, as there was much blood but no footprints. No motive was established, no suspects were ever arrested or even named, and the case remains unsolved to date.[3]

In a much less serious matter, nature served partial justice in the case of two carjackers in South Burlington, Vt. The *Free Press* described how the men stole a car and fled from the police. Five miles down the road, the men rounded a curve and crashed into a tree that had been blown down by the storm. The criminals managed to escape, but the damaged car was recovered.

The storm had long-term effects on insurance. Modern multiline homeowner's insurance, which covers a variety of perils in a single policy, was a nascent product in November 1950. Just a few months earlier, in September, the Insurance Company of North America introduced the first comprehensive homeowner's policy, which covered many hazards including fire, wind, lightning, theft, and riot, at a savings of around 20 percent over multiple policies.[4] Previously, homeowners typically bought fire insurance and optionally added endorsements for other perils. The new multiline policy proved wildly popular; a mere decade later (1960), more than $750 million in policies had been written.[5] These newer policies made it easier for purchasers to get loans, increasing homeownership.[6] The lack of such policies in November 1950, however, led to considerable headaches for homeowners in the Northeast.

Since the windstorm affected such a broad area, the effects on the insurance industry were immense. A few days after the storm, the National Board of Fire Underwriters hurriedly convened a special committee in New York City. The *Times* stated that the meeting was to prepare for a worst case scenario of a record number of claims. News reports in the following days confirmed the underwriters' fears as claims poured in. For example, in Troy, N.Y., the *Times-Record* noted that more than five thousand claims, approximately five times the previous record, were filed in the days following the storm. Vermont agents were overwhelmed with claims and the state insurance board anticipated record losses. Likewise, in New Hampshire, the storm was termed the largest catastrophic loss in insurance history for that state. The *Manchester (N.H.) Morning Union* then detailed how insurance companies set up special claims offices, purchased newspaper advertisements, and extended deadlines to help keep the workload manageable.

Many claims were denied because of exclusions. For example, homeowners with fire insurance but without extended coverage for perils like wind were out of luck entirely, something stated by insurance companies in advertisements and mentioned in newspaper editorials following the storm. In Vermont, one agent told the *Burlington Free Press* that only around 65 percent of his clients had wind coverage. Another area agent called householders without coverage "foolish" given that three windstorms had struck the state in the previous twelve years. Homeowners with wind insurance were not necessarily fully covered, either. Insurance companies did not pay for repairs from water seepage, even if it was caused by wind-driven rain that went under shingles or through walls. On the positive side, underinsured homeowners could deduct noncovered losses on their 1951 federal tax returns (provided that they had receipts substantiating their losses), as specified in the *New York Times*.

In New England, where insurance was underwritten by the New England Fire Insurance Rating Association, some homeowners with extended coverage were dismayed to learn that they were responsible for the first $50 in damage (approximately $500 today). Especially in southern New England, many homeowners had added wind coverage to their insurance following the 1938 Great New England Hurricane. However, a large number of trivial claims following the 1944 Great Atlantic Hurricane (and from routine weather in subsequent years) caused insurers to exclude losses of less than $50, according to the *Providence Journal*. In a similar article, the *Hartford Courant* noted that insurers added deductibles because many small claims were simply from normal wear and tear. Thus, $50 deductibles on homeowner policies began as an option in October 1948 and were made mandatory for policies written in September 1949 or later. Since policies were often written for a three- to five-year term, some homeowners were subject to the deductible and others weren't. Many of those subject to the deductible were unaware of the change in terms.

In 1950, state and local governments still bore the brunt of disaster response and costs. (The Federal Emergency Management Agency, FEMA, was not created until 1979.[7]) Federal response was limited to the financing of loans through the Reconstruction Finance Corporation, an agency created during the Great Depression, and from the loosening of lending restrictions on banks. The RFC proclaimed New England a disaster area, and the Boston office established a satellite office in Rockland, Maine, to help stricken lobstermen. This response was insufficient for Maine governor Frederick G. Payne, who complained to the *Boston Globe* that while loans were helpful, lobstermen needed direct relief to replace lost equipment and revenue. Farther south, federal loan requirements and terms were eased in New Jersey, Pennsylvania, and Delaware, thanks in part to intense lobbying by New Jersey U.S. House representative James C. Auchincloss. Still, the *Asbury Park Press* editorialized that direct relief should be provided, claiming that the federal government regularly "stole" tax revenues from New Jersey to spend on "politically favored" states in the South and West.

THE WINDSTORM THAT AFFECTED THE NORTHEASTERN UNITED STATES, WHILE severe, caused widely variable impacts. For the majority of residents, the windstorm was a nuisance, but those who did suffer damage often suffered severe catastrophic losses. The next chapter reviews a hazard that was less easily avoided: record cold in the South. Unlike with the wind, the impacts of this hazard varied considerably with race and class.

8

FRIGID

"November 1950 . . . will long be remembered by many citizens on account of the great drop in temperature on the 24th and 25th. The morning of the 25th was, over the entire State, colder than has ever been recorded in November."

—A. R. LONG, WEATHER BUREAU SECTION DIRECTOR, MONTGOMERY, ALA., NOVEMBER 1950[1]

AN OUTBREAK OF COLD AIR IS A MORE SUBTLE FORM OF DISASTER THAN A blizzard, flood, or windstorm. There are no mountains of snow or ice, rivers are not rampaging, and trees and other objects are not flying about. Cold weather slips in quietly, often gradually, until people suddenly notice its fierceness. While tornadoes, hurricanes, and other weather events may dominate the headlines, extreme cold and heat are the biggest weather-related killers in the United States. A recent study by the Centers for Disease Control and Prevention found that 63 percent of all weather-related deaths in the U.S., or around 1,260 persons annually, were from extreme cold.[2]

The exceptional cold weather that affected the South in late November 1950 was similarly deadly. The month began with unusual warmth, with daily and monthly records for high temperature. Mild conditions continued before progressively colder weather gradually arrived during the second half of November. But on November 23 and 24, much colder air began arriving, and the upcoming winter season suddenly bore down. Over the next few days, residents of the South experienced numerous inconveniences, from slippery roads to frozen engine blocks and pipes. Farmers suffered millions of dollars of losses in crops and equipment. Worst of all, dozens of impoverished residents, mostly African Americans, died in fires or from exposure. While the cold didn't cripple transportation and commerce like the snow and ice did elsewhere, the costs and loss of life associated with Superstorm 1950 were just as serious in the South.

MONDAY, NOVEMBER 20–THURSDAY, NOVEMBER 23

A COLD APPETIZER

After a relatively warm start, the weather turned much colder for the last third of November. A strong cold front swept through the region on the Monday and Tuesday before Thanksgiving (November 20–21). Strong winds—possibly tornadic—destroyed a home near Noxapater, Miss., injuring two, and three tornadoes were observed elsewhere in Mississippi and Kentucky. Following this front, the coldest air of the season, and in some places, the coldest air of the calendar year, flowed in. Inland locations experienced lows in the twenties, and northern Florida was quite cold, too, as Gainesville fell to 27°F and Tallahassee recorded its coldest weather in nearly a year (31°F). Only areas along the coast and far southern Florida remained above freezing. Nonetheless, while unusual for November, the cold was mild enough that damage to crops was minimal.

In Charleston, S.C., the cold indirectly caused the first death of the winter season. Eddie Meyers, approximately age 50, attempted to light an oil stove at about 8:10 p.m. on November 21. The stove exploded, engulfing his house in flames and burning him to death. Mr. Meyers was African American. Similar anecdotes were common in the following week.

Forecasters anticipated a return to seasonable, perhaps even mild, conditions for Thanksgiving. Longer-range forecasts hinted at another cold spell after the holiday, but there was not a consensus. Not that it mattered—weather forecasting in 1950 was still in its infancy, and forecasts beyond 24 hours were given little credibility anyway.

The weather for the midweek period was as expected, and Thanksgiving Day (November 23) was pleasant and warm, with temperatures above average. Late in the day, cold air arrived in western Tennessee. Forecasters in the South now agreed that a notable cold outbreak, one cold enough for snow, was likely, but collectively they underestimated the magnitude of what was to follow.

FRIDAY, NOVEMBER 24

WINTER ARRIVES

On Friday, November 24, many cities celebrated the beginning of the holiday season with parades and the arrival of Santa Claus. Ominously, winter arrived as well, accompanied by rapidly falling temperatures and snow as far south as central Mississippi and Alabama.

Similar to most cold waves in the South, this one progressed from northwest to southeast. Memphis, Tenn., was among the first cities affected. Rapidly arriving cold air and light snow caused a ten-car pileup on the Mississippi River bridge at 9:30 p.m. on Thanksgiving. Luckily just one person was seriously injured, but another similarly sized crash occurred again the following morning. The *Memphis Commercial Appeal* noted that these were the first snow-related crashes on the year-old bridge.

As the cold air advanced in the region, temperatures plummeted fiercely, with summer becoming winter in less than 24 hours. In Jackson, Miss., a balmy Thanksgiving afternoon with a temperature of 75 turned into a frigid night with a low of 22°F, a decrease of fifty-three degrees in fewer than 18 hours. The temperature that afternoon reached just 32°F, as recorded in the *Clarion Ledger*. Temperatures did not exceed freezing again until Saturday. In Tupelo, Miss., the *Journal* described how temperatures fell from a pleasant 67 on Thanksgiving to a chilly 18°F the following morning, with 3 inches of fresh snow to boot. The rare November snowfall triggered the postponement of the freshman football game between the University of Mississippi and Mississippi State. Meanwhile, the arctic blast continued to power southeastward. In Nashville, the *Tennessean* said that a "beautiful Thanksgiving Day turned into an icy nightmare" as the temperature rapidly fell and rain changed to snow around 9:00 p.m. A November-record 7.2 inches of snow accumulated in total. The story was similar 200 miles south in Birmingham, Ala., where the temperature fell fifty degrees in 18 hours, from 70 on Thanksgiving afternoon to 20°F by 9:30 the next morning. A November-record 1 inch of snow made for a miserable commute and crippled traffic heading north of town along steep, windy mountain roads. East in North Carolina, Asheville was buffeted by wind gusts as high as 47 mph, Winston-Salem experienced a thirty-nine-degree temperature drop in 9 hours, and Raleigh was hit by a sudden thunderstorm at around 8:00 p.m., as chronicled in local reports. In Walterboro, S.C., gusty winds felled a 165-foot ice-coated radio tower, knocking WALD off the air indefinitely, according to the *Charleston News and Courier*.

The sudden arrival of the cold air caused numerous traffic problems. Besides the chain-reaction crashes on the Mississippi River bridge in Memphis, more than eighty-two other crashes (and additional "minor mishaps" not worth reporting on) were listed from November 24 in the *Memphis Commercial Appeal*. Traffic problems were most common in mountainous areas and anyplace with sloped terrain. Local reports told how in Asheville, police set traffic lights to flash so motorists wouldn't need to stop on steep hills, while in Nashville, police took the extraordinary step of turning off some traffic lights. In Huntsville, Ala., police considered some steeply graded roads impassable and blocked them off, while Birmingham traffic moved along, albeit slowly. Police advised motorists to avoid mountain roads north of town unless they had chains. They also remarked that traffic was lighter than usual in several of these cities. The most severe crash occurred in Dalton, Ga., where a woman was unable to stop her car at a train crossing and skidded into the path of an oncoming train, as chronicled by the *Chattanooga Times*. Amazingly, although her car was demolished, the three occupants suffered only minor injuries, with a broken leg the worst.

Antifreeze and tire chains were hot sellers throughout the South. While a spate of below-freezing weather earlier in the month had encouraged residents of Tennessee to prepare, motorists in other Southern states were not ready and experienced frozen engine blocks and leaky radiators. Problems were most acute in places that rarely experience below-freezing weather, especially along an axis from Tallahassee, Fla., and Dothan, Ala., northeast into central South Carolina. Many stocks of antifreeze ran low, with one dealer in Tallahassee selling more than 600 gallons, according to the *Democrat*. In a widely syndicated report from Charleston, S.C., after a local meteorologist bought antifreeze, one enterprising dealer put up a sign advertising that "the weatherman just bought antifreeze—how about you?" Despite seventy-degree weather, he sold out within hours. Farther north, tire chains were hot sellers. One profiteer in Chattanooga, Tenn., marked up the price of his chains from $8 each to $10, as described in the *Times*. He sold his entire stock by noon.

Throughout the weekend, wrecker services remained busy with slide-off crashes, frozen radiators, and other automotive problems. But in multiple locations throughout the South, the police reported lighter than usual traffic volumes and relatively few crashes overall. They credited safe and prudent driving. In most places, traffic was able to move, though more slowly than usual. However, in the mountainous areas of northwestern North Carolina, accumulating snowfall and the exceptional cold made roads largely impassible even with chains. Many cars became stuck in ditches in Allegheny County, and more than a hundred cars became stranded in nearby Surry County, according to reports in the *Winston-Salem Journal*. Almost all driving deaths occurred in this region, as three people died in separate crashes on slick roadways. The only other weather-related fatal crash was near Savannah, Ga., where Ethel Blount was killed in a head-on collision. The driver of the other car had been blinded when a broken radiator hose sprayed water on his windshield, and the water instantly froze. The *Savannah News* mentioned that he was charged with reckless driving.

Radiator hoses were not the only pipes freezing. As the cold settled in on Friday and Saturday, frozen water and gas pipes became a widespread problem. Once again, the problems were most acute farther south, where below-freezing weather was less common. Heavy damage to water pipes resulted in numerous calls, according to the *Tallahassee Democrat*, while in Dothan, Ala., the *Eagle* portrayed how plumbers simply shut off the flow of water and scheduled appointments to replace broken pipes and valves at a later time. Often the branch lines from water mains to dwellings were the weak link; these were not protected by the warmth of the ground or insulated by the heat of the building. In Huntsville, Ala., more than three hundred calls were received by plumbers, often from rural areas that had recently received water service, as described by the *Huntsville Times*. Commercial building sprinkler systems also froze. At least three system failures occurred in Chattanooga, Tenn., and the *Atlanta Constitution* reported that at least nine systems failed there, the most for any city. In Florence, S.C., burst pipes ruined the printing presses of the *News*, making it impossible for the paper to publish. In what was truly an instance of bad luck, the pipes burst around noon on Sunday, the one time that employees were not normally in the building. Plumbers noted that breezy conditions exacerbated the cold conditions by pushing frigid air into crawlspaces, walls, and other unsealed areas. The breezes also cooled the pipes more quickly, similar to how people feel colder on windy winter days.[3] Problems with leaky pipes actually peaked a few days later as temperatures warmed and the pipes thawed.

Although city workers had to deal with frozen pipes and broken water mains in some places, gas and electric company employees were much busier dealing with record loads and outages. For example, near

Knoxville, Tenn., several families' Thanksgiving dinner was ruined by an ill-timed electric outage from 6:00 to 9:00 p.m., caused by unusually heavy usage. The outage was prolonged due to multiple circuit overloads as people ran heat on higher settings to warm up buildings quickly, according to the *News-Sentinel*. Shorter duration outages from overloads occurred in Nashville, Chattanooga, and other Tennessee cities as well. Outside of overloaded circuits, problems with the electric grid were minimal as the winds were not strong enough to bring down many power lines; this was true in areas that received snow also. Electric usage set a new record in Nashville. Gas usage was at record levels there as well as in Memphis, Atlanta, Charleston, S.C., and elsewhere.

Most gas and coal outages were caused by distribution issues, not shortages. Around 22 percent of homes in the South were heated with coal, and both coal and fuel oil delivery workers found themselves unable to meet the spike in demand.[4] Shortages from coal delivery problems occurred in Tallahassee, Huntsville, and Atlanta. In Atlanta, police took the unusual step of authorizing Sunday deliveries to help ease the coal shortages. Just under 30 percent of homes in Atlanta relied on coal.[5]

Natural gas delivery problems were more widespread. In Nashville, during the worst of the cold, the *Tennessean* outlined problems with low gas pressure caused by an overtaxed system. Officials assured customers that a new pipeline already under construction should be done in a few weeks—before the worst of winter. The *Jackson Clarion Ledger* told the story of how a gas line ruptured in Pascagoula, Miss., leaving fifty-five hundred families in that town and nearby Moss Point without heat on the coldest day in years. Farther east, water within natural gas lines froze and plugged numerous gas meters, halting the flow of gas. Problems were most severe in Savannah, where more than a thousand calls were received, according to the *News*. Charleston and Columbia, where three to four hundred calls per day were made over the weekend, also had extensive problems with frozen meters. Similar problems occurred in Chattanooga. Atlanta likely had problems too, but the problems with low gas pressure there likely overshadowed any problems with frozen meters. Luckily, Mother Nature fixed most of the problems as warmer temperatures thawed out the meters. (Ironically, though, the warmer temperatures reduced the need for gas heating.) Nonetheless, workers in all cities were kept busy with gas repairs. In Columbia, S.C., the *State* told how an additional eighteen workers were called in to help thaw out frozen meters. A normal weekend would have only four workers on duty.

The most serious natural gas problems occurred in Atlanta. Many customers complained of low pressure. Several automobile plants suspended operations as a result, and a city councilman threatened an investigation, claiming that the local gas provider was made aware of the problems several years ago. The utility told the *Constitution* that the extraordinarily low temperatures and customers adding gas appliances without notification were the causes of the shortage. Luckily the low gas pressure issues were more of an annoyance than an emergency.

SATURDAY, NOVEMBER 25–SUNDAY, NOVEMBER 26
COLDEST NOVEMBER WEEKEND EVER

Although the cold air arrived in most places on Friday, November 24, it peaked on November 25. As the *Charlotte Observer* termed it, November 25 was the day that "winter quit kidding around . . . and hit Charlotte with a record cold wave for this early in the year." The quote was appropriate for the entire South; even southern Florida was not spared. Tourists there joked that the Gold Coast should be renamed the "Cold Coast," and a group of young tourists from Brooklyn flew home because "it just couldn't be this cold up north" (though it was), according to the *Miami Herald*. While no records were set there on the twenty-fifth, record lows were recorded that morning almost everywhere else in the South. Most of these were all-time records for the month of November, and almost all are still standing more than seventy years later. Many places didn't warm much during the day, either, setting additional records for the coldest high temperature for the day, also known as "min max" records. A sample of records set in ten cities is shown in table 4.

TABLE 4. A sampling of November all-time cold records from November 25, 1950.*

Location	New November record low (°F)	New November record min max (°F)	Additional context
Nashville, Tenn.	−1	19	Earliest below zero temperature for the winter season.
Knoxville, Tenn.	5	14	Only single-digit reading on record in Nov.; only Nov. day that failed to reach 20°F.
Tupelo, Miss.	8	29	Only Nov. day on record that failed to exceed 32°F.
Huntsville, Ala.	1	19	No other record low in Nov. is below 10°F.
Montgomery, Ala.	13	30	Only Nov. day on record that failed to exceed 32°F.
Atlanta, Ga.	3	17	Record low is 11°F colder than second coldest Nov. temperature record.
Savannah, Ga.	15	32	Only Nov. day on record that failed to exceed 32°F.
Asheville, N.C.	1	8	Only Nov. day on record that failed to exceed 18°F.
Greenville, S.C.	11	25	1 of only 2 Nov. days on record that failed to exceed 32°F.
Pensacola, Fla.	22	40†	1 of only 2 Nov. days on record that failed to exceed 40°F.

*None of these records have been broken as of June 2022.
†Record for the date, but not for the month.

The coldest location in the entire region was atop Mt. Mitchell in western North Carolina, which at 6,684 feet is the highest location in eastern North America. Here the temperature fell to a bone-chilling −19°F.

THE SOUTH BURNS

The exceptional cold meant that heat requirements for many buildings were high. Additionally, with this being the first major arctic outbreak of the year, some heating systems required lighting for the first time. Unfortunately, the strain on heating systems caused many fires. These fires destroyed property, left people homeless, caused injury, and even caused more than a dozen deaths. The deadliest hazard associated with the cold outbreak in the South was fire.

In the South, many fires were caused by substandard equipment or unsafe heat sources. As temperatures fell, problems developed with stoves, stokers, and flues.[6] In Harriman, Tenn., a furnace caught fire around 4:30 a.m., destroying a general contractor's office, while in Knoxville, an overworked coal heater and defective flue caused significant damage at a residence—one of many such examples region-wide. In Winnesboro, S.C., an oil furnace exploded, causing a conflagration that destroyed the local theater. Luckily, the last show for the day had ended about 15 minutes before the explosion and fire, according to the *State* (Columbia, S.C.).

As the cold air continued to build into the weekend, deadly fires from heating systems became more widespread. Numerous other fires were caused by people attempting to thaw frozen pipes. All of these extra fires overwhelmed fire departments in many cities. In Memphis, the *Commercial Appeal* reported that more than a hundred calls happened from late Thursday (Thanksgiving) through the end of Friday, many from overheated furnaces, while in Nashville, more than twenty building fires caused $10,000 in damage. The *Nashville Tennessean* also noted that many other fires were not listed as the costs were negligible. A series of fires in Albertville, Ala., stretched the fire crews so thin that a house burned down while crews were busy fighting a larger fire in another part of town, as recounted in the *Birmingham News*. Many small fires caused damage in other Southern cities, including Columbia, Savannah, and Tallahassee. In Greenville, S.C., firefighters had an astounding forty-two calls in two and a half days. A forty-six-year veteran of the department told the *Greenville News* that he couldn't recall a previous instance of more than ten calls in one day.

Several spectacular fires destroyed large structures. A malfunctioning stoker furnace caused $2,000–$3,000 in damage at a hospital in Florence, S.C., while in Raleigh, N.C., the entire third floor of a women's dormitory at St. Augustine's College was gutted by a chimney fire. Fortunately, only around forty-five students were in residence over Thanksgiving break (approximately seventy students lived in the dormitory), and all students were able to escape without injury. Unfortunately, nearly all personal items on the third floor were destroyed. In Montgomery, Ala., a fire that began in the chimney of a coal heater destroyed a wooden prison barracks that housed just under three hundred prisoners. The *Huntsville Times* reported that when discovered, the fire was considered small enough that it could have been doused "with a bucket of water," but guards couldn't obtain a bucket of water since the water pipes were frozen. In Asheville, N.C., Marion High School suffered heavy damage from fire, and the Macedonia Baptist Church (a small neighborhood church) was destroyed, as described by the *Citizen*. One of the most spectacular weather-related fires occurred in Tupelo, Miss. The large First Baptist Church and adjacent Education Building were destroyed in a 15-hour fire late on the night of November 24. Causing an estimated $250,000 damage in 1950 dollars (more than $2.5 million today), the fire was sparked in the afternoon when someone tried to thaw frozen pipes in the men's restroom by burning scraps of paper, according to the Jackson *Clarion Ledger*. Although the fire department put out this first fire, apparently some embers lodged in the walls and smoldered for hours before the main conflagration was noticed around 8:00 p.m. Tupelo firefighters were hampered by single-digit temperatures and a 10 mph wind, though thick snow on the roofs of neighboring homes prevented their destruction.

Already stretched thin by many calls, the firefighters were further hampered by both bitter cold and breezy conditions. The *Atlanta Journal* told how in Cobb County, Ga., water in a tanker truck froze in the single-digit temperatures, causing two homes to be damaged by fire, one of which was destroyed. In Augusta, Ga., seventeen-degree weather hampered the response to a $200,000 fire at the Southern Cotton Oil Company, while windy conditions in Moultrie, Ga., helped fuel a $2.25 million warehouse fire started by spontaneous combustion, according to the *Charleston News and Courier*. In Gastonia, N.C., a downtown department store and two other buildings were destroyed by a large fire; damage was estimated at more than $250,000. Firefighters had extreme difficulty working as temperatures near zero froze water into sheets of ice on the streets and on their uniforms.

Some fires were simple accidents with deadly consequences. While around half of homes in the U.S. had central heating in 1950, in the South only 22 percent did.[7] This meant that most homes were heated by stoves of some sort. Because heating units were in occupied rooms instead of basements or garages and regular tending was required, this made them much more hazardous. Near Holly Springs, Miss., four people, including three children, died when a man threw kerosene on a small fire inside a stove, according to the *Jackson Clarion Ledger*. The *Chattanooga Times* recounted how Anderson Lowe was burned critically while kindling a stove. Sadly, at least half a dozen deaths were caused by people's clothing catching fire, including Barbara Coleman, age 12, whose dress caught fire in Gainesboro, Tenn.; Jane Keener, 78, in Albertville, Ala.; and several wheelchair-bound seniors who burned to death in their chairs in Georgia and Alabama. In Gadsden, Ala., Patricia Leath went to a neighbor's to borrow a knife and returned home to find her house ablaze from a coal stove that had exploded. Her children, ages 4 and 2, did not survive. In Millry, Ala., a fast-spreading fire killed Sallie Strickland, 85, while her daughter was in the backyard hanging clothes. The wheelchair-bound Strickland was unable to escape, and her daughter was unable to penetrate the flames and smoke to rescue her, according to accounts in the *Birmingham News*. In total, fatal fires killed at least twenty people in the South, including at least eight children and one firefighter.

Death from exposure also occurred most often in the South. In Chattanooga, the *Times* reported that Mary Ann Jackson, 69, was found dead in her frigid room. Meanwhile, the *Johnson City (Tenn.) Press-Chronicle* told how Betty Watson, 79, died from exposure after being found in a ramshackle building with no water, electricity, or heat. Ms. Watson's death was particularly tragic as the family had no income since her son-in-law had been killed in a mining accident "some time ago." Neighbors had recently informed Community Chest and the Salvation Army about the family's destitute condition, but apparently no action had been taken. Similar conditions led to the death of Reece Smith Jr., 35, who was found dead in his "shack" near Erwin, Tenn., and Charlie Baker, an "aged Negro," who was found dead in a house in Adel, Ga., as mentioned in reports from Asheville and Savannah. At least another dozen exposure deaths occurred, with reports found in Memphis, Nashville, Atlanta, Columbia, and Birmingham newspapers; many of these deaths occurred outdoors. Few details were available because the victims died alone. Detail may also have been lacking as the dead were primarily African Americans.

Race was a key determinant in who died from fire or exposure in the South; most victims were African Americans. As discussed earlier, African Americans in the South were treated as second-class citizens in 1950 and had few rights and even fewer opportunities.[8] The squalor in which many African Americans lived almost surely played a role, as faulty heating systems—or no heat at all—and substandard living conditions contributed to many deaths. The Census Bureau began keeping track of poverty in 1959, and at that time, a majority (55.1 percent) of African Americans nationwide lived in poverty, compared with 18.1 percent of whites. But these numbers mask how much poorer

Southerners—of all races—were. In 1959, the region-wide poverty rate was 40 percent (led by Mississippi with 54.1 percent of all citizens in poverty), and the ten states with the highest poverty rates in the country were all in the South. Poverty rates for African Americans were even higher, with statewide poverty rates between 60 and 85 percent.[9] While these numbers are from 1959, the 1950s was a period with large economic growth. Poverty rates steadily declined in the 1960s; this trend probably began in the 1950s. In addition, a recession occurred in 1949, and presumably increased poverty in 1950. Thus, it is not surprising that the South led the nation in deaths from fires and exposure associated with the storm, or that African Americans, who were the most economically disadvantaged group, were also the most affected group.

A COLD DAY FOR FOOTBALL

On a much lighter note, Saturday, November 25, was one of final days of the college and high school football seasons. Many rivalry games with high stakes for bowl game assignments—or simply bragging rights—were scheduled. Despite the cold, football games in the Deep South largely happened as planned. The primary effect of the cold weather was discomfort for players and spectators; the games themselves were largely routine in comparison to the blizzard and mud bowls elsewhere. In Jacksonville, Fla., eighteen thousand fans shivered through sunny but cold weather (game time temperature approximately 32°F) to watch Alabama defeat Florida, 41–13. The *Florida Times Union* of Jacksonville estimated that several thousand fans failed to show up, but it provided no evidence to support this claim. In Columbia, S.C., Wake Forest overcame twenty-four-degree weather and a frozen field to defeat the University of South Carolina by a score of 14–7, before what the *Raleigh News and Observer* termed ten thousand "hardy souls."

In northern portions of the South, snow affected the games, and colder temperatures had more serious effects. In both Tennessee and North Carolina, snow caused many high school football games to be canceled or postponed, though most of them were exhibition contests anyway. In college football, the cold had deleterious effects. In front of forty thousand shivering fans, Duke defeated the University of North Carolina 7–0 in an ugly game featuring six turnovers. The only score of the game came on a short field, after Duke blocked a North Carolina punt.

The biggest effect of the storm was on the Kentucky–Tennessee game in Knoxville. In a fierce rivalry game, the third-ranked University of Kentucky football team sported an undefeated record and a strongly rumored (but officially unconfirmed) Sugar Bowl invitation. Though the stadium was filled with 7 inches of snow, and more than 100 tons were on a tarp covering the field, there was no public discussion of postponing the game. Instead, more than eighty workers spent in excess of 7 hours clearing what the *Knoxville News Sentinel* estimated to be a thousand wheelbarrow loads of snow. Workers also cleared the stands sufficiently for patrons to find their seats, but most other snow in the stands was left in place. Despite the frigid weather, poor road conditions, and snow-covered seats, an estimated forty-five thousand of fifty-two thousand ticketed fans attended the 2:00 p.m. contest.

Though the weather conditions were fair with no precipitation falling (unlike the Ohio State game 350 miles to the north), the frigid temperatures caused an equally ugly show of offensive ineptitude. Kentucky and Tennessee fumbled the ball an incredible sixteen times, losing twelve of those fumbles to the other team, and Kentucky threw four interceptions for good measure. Like the Duke–North Carolina game, the only score came off a turnover, as Tennessee recovered a Kentucky fumble and completed a touchdown pass soon after. The Kentucky Wildcats looked nothing like the third-ranked football team in the country, as they spent all but two drives on their side of the field. A sports writer for the *Nashville Tennessean* remarked that "the wonder of the game is the score. No other football team ever committed the mistakes Kentucky did yesterday and escaped with a 7 to 0 licking."

ANOTHER COLD MORNING ON SUNDAY

Conditions on Sunday morning, November 26, were frigid once again. While the cold had eased slightly in western portions, record lows occurred throughout eastern Tennessee, Alabama, Georgia, and the Carolinas (see table 5). Florida bore the brunt of the cold on Sunday with records set from Pensacola (30°F) in the north all the way south to Key West (56°F). All-time November record lows were set in Fort Myers (34°F) and West Palm Beach (36°F). While these numbers may not seem cold to people who regularly experience snow, such temperatures in South Florida are quite uncomfortable given the lack of building insulation and heating systems. Many Southern cities again set records for the coldest high temperature (min max) for the day. Most records set this day were not records for the month, but this was simply because the previous day had been colder.

TABLE 5. A sampling of cities that set new records on November 26, 1950.[*]

State	Cities that set both new record lows and record cold highs (min maxes)	Select cities that set new record lows
Tennessee	Chattanooga, Knoxville	—
North Carolina	Asheville, Greensboro	Charlotte,[†] New Bern, Raleigh, Wilmington[†]
South Carolina	Florence,[†] Greenville[†]	Charleston, Columbia
Alabama	—	Birmingham, Huntsville, Montgomery, Mobile
Georgia	—	Atlanta, Macon, Savannah
Florida	Gainesville, Melbourne, Tampa, West Palm Beach[†]	Fort Myers,[†] Key West, Orlando, Pensacola, Tallahassee

[*]Most of the records have not been broken, with the exception of some in the Carolinas broken in Nov. 1970.
[†]Morning low temperature was also an all-time record for Nov.

As temperatures began warming slightly, burst pipes in many places started to thaw, causing new problems with water leaks throughout the region. In Florence, S.C., the *News* reported that more than two hundred repair requests on Sunday and Monday caused plumbers to warn residents that a complete fix may take seven to ten days. In Charleston, S.C., plumbers worked "day and night" dealing with leaks. Leaks were so widespread that the city's water pressure decreased to a level where the American Tobacco Company had to suspend operations. Near Asheville, N.C., expansion and contraction from the dramatic temperature changes broke several water mains, as listed in the *Citizen*.

THE EFFECTS ON AGRICULTURE

On Sunday, farmers also began assessing the damage to their crops and equipment. Outside of deadly fires, the most serious impacts of the storm in the South were on agriculture. Millions of dollars in crops were lost, primarily in Florida. Millions more dollars were lost when water in radiators and engine blocks froze; water also froze in irrigation systems, causing yet more losses. However, there were also some benefits, especially in the Cotton Belt, where many pests were killed by the bitter cold.

Crops in Mississippi, Tennessee, and northern portions of Alabama and Georgia were somewhat protected by snow cover and the fact that few crops are grown in late November.[10] But farther south, widespread crop losses occurred in northern Florida and nearby adjacent areas of Alabama and Georgia. Near Augusta, Ga., more than 80 to 100 percent of vegetables, such as turnips, collards, and rutabagas, were ruined, along with the majority of oats and other grains. Blue lupine, an important wintertime soil-building crop and one that provided seeds that could be sold for five cents per pound, was a total loss. Initial estimates of crop losses in Georgia and South Carolina were around $10–$15 million, equivalent to $100–$150 million today; these estimates excluded widespread damage to tractor engine blocks and water systems from water freezing. The narrative in southern Alabama was similar, with blue lupine and farm equipment the loss leaders in the Dothan area, as reported in the *Eagle*. Southward in Florida, poinsettias and other ornamentals in and around Jacksonville were ruined, according to the *Florida Times Union*, and more than $30,000 ($300,000 in modern dollars) in ferns were destroyed in Volusia County, northeast of Orlando.

Despite these losses, farmers in the Citrus Belt (centered around Orlando) and elsewhere on the Florida peninsula breathed a sigh of relief. Area farm extension agents had warned of a possible hard freeze just prior to Thanksgiving. A hard freeze occurs when temperatures fall to or below 28°F for a specific period of time. Smaller fruits like tangerines will be adversely affected after just a few hours, while it will take longer for oranges and longest of all for grapefruits to suffer ill effects. A hard freeze could have triggered a citrus embargo and cost farmers millions of dollars. Fortunately, the cold air was moderated by warming breezes off the Gulf of Mexico, thanks to the rapidly intensifying storm near Lake Erie. Although record low temperatures were set in many locations, the readings were warmer than forecasted. For example, the *Florida Times Union* described how forecasters expected a low of 16°F for Jacksonville, but the actual low was 26°F. While still record cold, clearly it was a much better outcome. (Interestingly, in most other Southern states, low temperatures were *colder* than forecasted.) As a result, the Florida Citrus Commission canceled an emergency meeting tentatively planned for November 27.

The celebrations were premature, however. On subsequent nights the winds died down, and the lingering cold air killed many crops. Central and southern Florida, especially around Lake Okeechobee, were hardest hit. Near Orlando, tangerines, whose small size makes them more vulnerable to cold, suffered "considerable" loss, and cabbages and vegetables suffered extensive losses as well.[11] Farther south toward Wauchula and Arcadia, more than 90 percent of the vegetable crop was ruined, the *Miami Herald* reported. Multiple other reports listed additional devastation: more than 90 percent of beans growing on 35,000 acres near the Everglades were lost, costing an estimated $10 million 1950 dollars; cold air also damaged countless potatoes. In sum, while most citrus crops avoided serious damage, cold air destroyed most tender crops. The cold outbreak was the final chapter in a miserable fall season for Florida farmers. An unusual late-season hurricane had caused widespread damage in October, and then November was extremely dry, with a record cold wave to close out the month.

In cotton-growing areas of the South, the cold spell was welcomed by weary farmers tired of fighting pests. To be sure there were crop losses, especially oats.[12] But protection from snow cover helped, and not a lot of crops are grown in Mississippi and Alabama that time of year.

Beneficially, the cold caused large-scale damage to boll weevils and other pests. An invasive beetle that migrated to the U.S. in the late nineteenth century, the boll weevil devastated the region's cotton economy in the early twentieth century and continued to cause deep economic losses despite widespread use of insecticides.[13] In 1950 alone, the boll weevil was blamed for cotton losses of $600 million ($6 billion in modern dollars), as it destroyed more than 28 percent of the cotton crop in Alabama and more than 16.5 percent of the cotton crop nationwide. At the time, this loss was the second greatest on record, behind only 1949 in terms of dollar amounts. However, sustained temperatures far below freezing killed many boll weevils in Tennessee and northern parts of Mississippi, Alabama, Georgia, and South Carolina. Area entomologists all expressed optimism for the 1951 crop. One told the *Greenville (S.C.) News* that he anticipated that the cold might set the boll weevil back two years. Other pests, such as the screwworm and sweet potato weevil, suffered heavy losses as well, especially in areas where temperatures fell below 10°F. Fruit trees also benefited from the cold. Peach and other fruit trees need a certain number of chill hours to bear fruit (chill hours are those when air temperatures drop below 45°F). Insufficient chill hours in the previous two winter seasons had reduced the size of the fruit crop, according to the *Jackson Clarion Ledger*.

MONDAY, NOVEMBER 27–FRIDAY, DECEMBER 1
A LONG, SLOW CLIMB

Monday, November 27, brought a new workweek and new problems. Schools sought to resume classes, and workers expected to return to work following the extended holiday weekend. Although the cold had begun to moderate slightly, snow and ice kept roads in eastern Tennessee, western North Carolina, and northern Alabama treacherous, and localized utility and water problems caused problems elsewhere.

Schools were initially unaffected by the cold since they were closed for the Thanksgiving holiday and extended weekend. In places where snow did not fall, most schools reopened as scheduled, save for the occasional school closed due to heating problems or frozen pipes. For example, the *Asheville Citizen-Times* told how a school in Swannanoa, N.C., was closed because of a defective stoker, while the *Savannah Morning News* noted that a local school was closed because the heating facilities were "out of order." Conditions were not optimal for learning, though. In Southern areas, like Dade County, Fla., students shivered through classes in buildings with insufficient (or no) heating systems, according to the *Miami Herald*. Colleges opened on schedule, except for Asheville Biltmore College, which remained closed because of its location on a steep, ice-covered road. Ironically, when road conditions improved enough for the college to reopen on Wednesday, November 29, heating problems led to its closure within an hour.

Lingering snow cover, persistent flurries, and continued breezy conditions extended the holiday for most schoolchildren in eastern Tennessee and western North Carolina. In Tennessee, schools were closed in many counties, primarily in the eastern third of the state, including populous Knoxville. The *News-Sentinel* mentioned that city-operated schools reopened on Tuesday, but those in Knox County remained closed until Wednesday. Most schools located in the hillier surrounding counties did not reopen until the following week. It was a similar story in western North Carolina, where cold weather and snow-covered roads extended the break for nearly all students located north or west of Winston-Salem. While city schools in Winston-Salem managed to open on time, those in surrounding Forsyth County remained closed for an additional day because of poor road conditions, frozen pipes in some buildings, and a teacher shortage resulting from employees being stuck out of town, according to the *Journal*. Superintendents in other counties hoped to open later in the week, but as the days went on they became more pessimistic. By Wednesday, November 29, Ashe County Schools (in the extreme northwestern part of North Carolina) announced to the *Winston-Salem Journal* that it planned to remain closed until the following Monday, December 4. Adjacent Watauga County was slower to cancel, but on the following day it, along with Alleghany County (east of Ashe County), also announced closure for the rest of the week, the result of "huge drifts" blocking secondary roads.

Southwest in Asheville, the story was similar. Asheville City Schools delayed opening by one day and surrounding Buncombe County Schools delayed two days, according to reports in the *Citizen*. These closings caused a great complication for teachers, however, as North Carolina law mandated that teachers must work twenty days in a pay period to get a paycheck. Because of the closing and the upcoming Christmas holiday, teachers were short several days, meaning that instead of getting paid just before Christmas, they would not receive another paycheck until January. (Remember, in 1950 credit cards as we know them didn't exist.[14]) Although the superintendents of Buncombe and seventeen other western counties immediately applied for relief, the State Board of Education was not scheduled to meet until Thursday, December 7. Thus, Buncombe County Schools took the extraordinary step of holding school on Saturday, December 2, and warned parents to expect school on the following Saturday also. Fortunately, on December 7, the state superintendent of education waived the payment rule. Schools were permitted to pay teachers in December despite them not having worked the requisite number of days. The superintendent sternly cautioned schools to make up the missed days as early as possible in 1951, as quoted in the *Asheville Citizen*.

Commerce, generally speaking, resumed normal activities on Monday. Some businesses closed because of problems with water or gas pressure, or as a result of fire, but these problems were isolated in nature. Crop markets were one of the rare businesses affected, as people had to travel considerable distances to them. In both Lexington, Ky., and Nashville, Tenn., after initially trying to conduct business as usual, market owners delayed tobacco sales from Wednesday to Thursday. Bad weather in the eight-state tobacco region had hampered both deliveries of tobacco and buyers' arrival. Due to the freeze in Florida, vegetable prices spiked over the weekend, but they gradually fell to normal values as the week went on.

As the week began, several stories of rescue from the holiday weekend emerged. In Indianola, Miss., a hunter was lost for almost 24 hours in weather as cold as 16°F. He was found on Saturday, cold but in good health after sustaining himself with a fire overnight, according to the *Greenville (Miss.) Delta-Democrat Times*. East in the Carolinas, the Coast Guard rescued twenty-one stranded boys from islands near Charleston, S.C. Further north, Captain Charlie Mack Williams, 60, of the Hatteras–Ocracoke mail boat was swept overboard. Two bystanders pulled him to safety from the rough seas. Near Mt. Mitchell, N.C., 29-year-old Mildred Harkey and her 7-year-old son, Allan, drove past a "Closed" sign and became stuck on the Blue Ridge Parkway. In 1950, the parkway was not maintained for traffic use in snowy conditions, though it was occasionally cleared of snow to protect the road surface. After nearly 18 hours in near-zero temperatures, a passing plow driver spotted their car and authorities staged a rescue. By slowly burning car upholstery piece by piece, the Harkeys had stayed alive. Nonetheless, both were admitted to the hospital, and medical personnel expected it to be "some time" before Ms. Harkey could walk again.

Level heads saved more than thirty people in two separate incidents from carbon monoxide poisoning.[15] The *Atlanta Journal* depicted one incident. A group of twenty-eight Georgia high school students traveling to the University of Georgia–Furman football game were woozy at a gas stop. Authorities discovered that the exhaust pipe on the brand new bus was too short, which allowed exhaust gases to enter the tightly closed bus. Luckily, only one student required hospitalization. In a UPI Wire story, the Purkey family was nearly killed by carbon monoxide while on their way to a holiday parade in Knoxville, Tenn. Robert Purkey Sr. was driving, and his wife became concerned when their 2-year-old son, riding in her arms, turned blue and became unresponsive. She then noticed that their 5-year-old son, Bobby, was passed out in the back. The parents immediately stopped the vehicle and resuscitated Bobby, preventing tragedy, though both boys needed supplemental oxygen. Coincidentally, the parade the Purkeys planned to attend had been postponed because of the snow.

Clear thinking also saved a 10-year-old girl from certain death in Aiken, S.C. When her house caught fire she was unable to get downstairs to exit. Instead, she climbed out of her upstairs bedroom window and calmly waited on the roof for rescue. The house was a total loss, but the girl was saved, earning praise from the local fire chief for her composure. The *State* (Columbia, S.C.) reported that the cause of the fire was undetermined.

Another tragic but ultimately positive story emerged from Knoxville, Tenn., and the *News Sentinel*. On Tuesday, November 28, Mary Massengill filed a restraining order and divorce papers against her husband of twenty-seven years, Gilbert Massengill. A child bride, she had been married at age 13 to a man twenty-two years her senior, and in the court filing she documented numerous examples of horrific abuse. The triggering event occurred on Sunday, November 26, when her husband pointed a loaded gun at her and forced her out of the house in subfreezing weather. After trudging several miles through snow to a neighbor's house, she finally had the determination to end her marriage.

THE COLD OUTBREAK IN THE SOUTH WAS NOT JUST RECORD-SETTING — IT established new limits for what people considered extreme November cold. The fires and deaths from exposure showed how the abject poverty and substandard standard of living of African Americans increased the number of weather-related deaths. The agriculture industry, while celebrating the deaths of pests, dealt with gigantic losses in the millions of dollars resulting from the sharp freeze. Frozen pipes and radiators caused great inconvenience for many residents.

The cold also brought out the best in people. Besides the stories of rescue, citizens in several areas responded to emergency pleas for help from charitable organizations. In Asheville, N.C., for example, the appropriately named *Citizen* described how local citizens overwhelmed the local Salvation Army office with donations of blankets, food, and coal. Many news accounts were filled with pictures of kids sledding and playing in the snow (though some kids suffered serious injury in the process). Hundreds of kids still watched Santa arrive in cities across the region, their joy enhanced by the Christmas-like weather he brought to places like Tallahassee, Tupelo, and Birmingham (though not Knoxville, where there was too much snow for the parade). Newspaper editorials remarked on the beauty of winter, celebrated the community spirit brought out by the cold, and expressed thankfulness for being spared the chaos of snowbound places. In small doses, perhaps the winter cold was not so bad.

PART 3

THE UPSHOT

9

THE MODELERS

"The advent of the large-scale electronic computer has given a profound stimulus to the science of meteorology. For the first time the meteorologist possesses a mathematical apparatus capable of dealing with the large number of parameters required for determining the state of the atmosphere and of solving the nonlinear equations governing its motion."
—JULE CHARNEY, APRIL 27, 1955[1]

MARK TWAIN IS OFTEN CREDITED WITH COINING THE PHRASE "EVERYONE talks about the weather, but nobody does anything about it." Surely everyone was talking about Superstorm 1950 in the days immediately following due to its record-setting snow, ice, flooding, wind, and cold felt in a large portion of the country. Few people, though, could do anything about the weather itself.

However, one group of people affected by the storm was able to do something about prediction of the weather. In Princeton, N.J., a legendary group of meteorologists, including Jule Charney and Norm Phillips, and the mathematician Jon von Neumann were affected by the storm. Winds at nearby Trenton gusted over 100 mph, caused extensive damage, and knocked out power to much of the town and university.[2] As surprised as anyone else by the storm, these young men (they were not yet legends, but well on their way) saw an opportunity to improve scientists' understanding of meteorology. They had recently come together at Princeton to develop a computer that could simulate, or model, the atmosphere. As they were already excited by previous success with a crude computer simulation, this badly forecasted storm provided a perfect case study to test their new theories about the dynamics of the atmosphere.

The unusual movement of the storm also made it ideal for creating models that forecast storm surge. The winds were both record-setting and from unusual directions. A storm surge model able to correctly forecast surge in connection with Superstorm 1950, as well as more common nor'easter storms, would have greater accuracy.

EARLY ATTEMPTS AT MODELING

Like maps or physical models of objects, meteorological models are simplified representations of the atmosphere. The vector calculus and differential equations that form the backbone of meteorology are mathematically complex and difficult to solve without a computer. Furthermore, extensive amounts of weather data are required. Modern computer models perform 2.8 quadrillion calculations *every second*.

The first meteorologist to attempt modeling was the British scientist Lewis Fry Richardson, long before the computer age. A pacifist Quaker, Richardson served in World War I as an ambulance driver. He would transport injured soldiers by day and make meteorological calculations at night. In the thick of the battle of Champagne, Richardson lost his papers for a time. Luckily, they were found in a coal heap several months later.[3] Richardson created a model (on paper) with data from several days in May 1910 when observatories across Europe had coordinated the launch of weather balloons and kites. His model would simply make a 6-hour forecast of the weather conditions on May 20, 1910.[4]

Richardson finished his work after the war ended and in 1922 published his results as a book titled *Weather Prediction by Numerical Process*. Although Richardson's model failed (it produced absurd results), with additional calculating power he felt that such prediction would be possible. Richardson even envisioned a "forecast factory" where sixty-four thousand workers would produce calculations that could be used to create weather forecasts.[5] Richardson goes on to describe it thusly:

> Imagine a large hall like a theatre, except that the circles and galleries go right round through the space usually occupied by the stage. The walls of this chamber are painted to form a map of the globe. The ceiling represents the north polar regions, England is in the gallery, the Tropics in the upper circle, Australia on the dress circle and the Antarctic in the pit. A myriad [of] computers are at work upon the weather of the part of the map where each sits, but each computer attends only to one equation or part of an equation. The work of each region is coordinated by an official of higher rank. Numerous little "night signs" display the instantaneous values so that neighbouring computers can read them. Each number is thus displayed in three adjacent zones so as to maintain communication to the north and south on the map. From the floor of the pit a tall pillar rises to half the height of the hall. It carries a large pulpit on its top. In this sits the man in charge of the whole theatre; he is surrounded by several assistants and messengers. One of his duties is to maintain a uniform speed of progress in all parts of the globe. In this respect he is like the conductor of an orchestra in which the instruments are slide-rules and calculating machines. But instead of waving a baton he turns a beam of rosy light upon any region that is running ahead of the rest, and a beam of blue light upon those who are behindhand.[6]

Concerned about the military using his research, Richardson soon halted his work in meteorology. Close to twenty-five years later, in 1946, a group of meteorologists met at the Institute for Advanced Study (IAS) in Princeton, N.J. Young and open to revolutionary ideas in the way that young people often are, they discussed the idea of using a computer to help solve meteorological equations of the atmosphere. This idea was radical at the time; the problem was viewed as a great mathematical challenge with no guarantee of success. Many meteorologists, including American Meteorological Society president Henry G. Houghton himself, expressed great doubts.[7]

Nonetheless, the Princeton group, which included future notables such as Jule Charney, John von Neumann, Jerome Namias, and Norm Phillips, pressed on. In 1947, Charney became more widely known when the *Journal of Meteorology* devoted an entire issue to publishing his doctoral thesis.[8] It was a breakthrough in meteorology; Charney had discovered critical equations that help govern storm formation and development. Unfortunately, the math was too cumbersome and challenging to solve by hand. However, computers could potentially solve the equations and allow for his theories to be tested and refined. Charney was selected to lead the IAS's Meteorology Project. He immediately began advocating for development of computer models.[9]

Computer models, thus, were initially conceived as mathematical tools to help refine scientists' understanding of the atmosphere. In contrast to modern computer models, the primary purpose was not forecasting. Instead, these early computer models would "forecast" past weather events. The results could then be compared with the actual weather events to assess the validity of the mathematical equations and solutions. Naturally, detailed weather information from past events would be needed for such models.

The Princeton group secured funding from the U.S. Navy to build a computer for its purposes at Princeton. Development would take several years. In the meantime, several members ventured to Aberdeen, Md., in March 1950 to attempt to forecast the weather using the army's Electronic Numerical Integrator and Computer (ENIAC).

FIGURE 9. Glenn A. Beck and Betty Snyder program ENIAC. (U.S. Army photo.)

Finished in 1945, ENIAC (see fig. 9) filled 1,800 square feet—a space nearly twice the size of a typical 1940s house[10]—and consisted of 17,468 vacuum tubes, 70,000 resistors, and half a million solder joints. Programming the machine was quite tedious. First, each calculation needed to be broken into many small steps; the process of ascertaining these could easily take a month or more. Once the process was determined, setting up thousands of switches to operate it often took a day or longer.[11] And it was unlikely that the machine would work the first time. On average, ENIAC would run for 5 to 6 hours before something went wrong. Debugging a program could take days or weeks.[12]

Despite all these complications, ENIAC was a thousand times faster than the best numerical calculator, and it could perform calculations in 30 seconds that would take a human with a calculator 20 hours to complete.[13] (A modern computer could do similar computations in mere milliseconds.) The sheer power of ENIAC was unequaled at the time, and despite the hassles, the opportunity to use it was too valuable to refuse. After working around the clock for thirty-three days and generating more than one hundred thousand punch cards, the IAS team was able to successfully make four 24-hour forecasts that resembled the actual weather event they were trying to simulate.[14] The scientists were pleasantly surprised by the quality of the forecasts and excited for the future.

SUPERSTORM 1950 ENTERS THE PICTURE

Seven months after the successful ENIAC experiment, Superstorm 1950 occurred. An extreme event that was badly forecasted, the storm, with its unusual nature, presented an inspiration for the modelers. If their computers could predict this storm, with its unusual track and weather conditions, surely they could predict simpler, less complex weather systems. This storm would become a critical case study for modelers, and one to be regularly used as a test case for the next fifty years.

Norm Phillips, another member of the Princeton group, developed a simple two-layer equation model to expand on Charney's earlier work. (The term *layer* refers to the number of horizontal surfaces in a model. For example, a four-layer model might calculate weather parameters at ground level and also at 1 mile, 3 miles, and 5 miles above the ground.) In 1951, Phillips tested his model by hand calculating values for Superstorm 1950, subsequently publishing his results in the *Journal of Meteorology*.[15] His results were mixed, with Phillips himself remarking on the need to add a third layer and change some assumptions. This was the first of many experiments involving Superstorm 1950.

By 1952, the Princeton computer was running. It was seven years newer than ENIAC and had vastly superior capabilities. Charney remarked that it could complete in 5 minutes a task that had taken 24 hours with ENIAC.[16] Nonetheless it was still cumbersome and difficult to use. There were no computer programming languages. Instead, a staff of eight coders would convert mathematical and logical instructions into a code the machine could read.[17] A large group of engineers was needed as well. Access to the computer was very limited, and described by Bruce Gilchrist, one of the programmers who aided Charney.

> The typical routine was for the engineering groups to have the machine all day for maintenance and development activities. Then the second and third shifts would, if we were lucky, be available to users like me. Needless to say, I pulled a lot of all-nighters![18]

With the model running, a positive feedback loop developed, where equations could be developed, tested on the model, and then refined and tested again. Further rapid advances in computing power also helped. For example, in late 1953, additional tests were run on an IBM 701, which had more memory, improved input/output, and tape drives. This allowed the modelers to create a weather simulation with five vertical layers.[19]

Throughout this time, Superstorm 1950 was repeatedly used as a test case. In 1953, Charney and Phillips published a paper outlining how a two-and-a-half-dimensional model had outperformed a two-dimensional model for this storm.[20] One year later, Charney's three-layer model provided further improvement in forecasting the magnitude of the storm, though the location was incorrect.[21] In 1958, Phillips used modified versions of the equations with a stream function to see if the model's accuracy could be increased, again with Superstorm 1950 as a test case. The experiment was successful; the two-layer model with a focus on wind fields rather than temperatures was a vast improvement over conventional geostrophic models.[22]

The continued success of the modelers attracted wide recognition, with Charney publishing his simulated forecasts of Superstorm 1950 in the *Proceedings of the National Academy of Sciences*.[23] Charney even sent one of his papers to L. F. Richardson, then in his 70s, and received a congratulative and supportive letter in response.[24] Reflecting years later in 2004, Phillips directly attributed Charney's (and his) work with the storm as leading to the creation of the Joint Numerical Weather Prediction Unit in 1954. A collaboration between the Weather Bureau, air force, and navy, the JNWPU issued its first forecasts in 1955, using Charney's three-dimensional model.[25] Although those forecasts were not of usable quality, by 1958 computer models were able to successfully forecast future weather. Great strides in modeling and forecast quality have been made since then.[26] The development of meteorological modeling could be an entire book itself.[27]

WAS SUPERSTORM 1950 PREDICTABLE?

No computer models capable of forecasting existed in 1950, but would modern models have been able to forecast the storm? The answer is a qualified yes: Although the storm was unusual, it was predictable with modern techniques.[28] However, even if forecasters had modern models then, advance notice of the storm would have been limited to about three days due to a lack of upper level data. In the 1950s, the upper level observing network was in its infancy, which means that upper level reports were not made with the same consistency and data density as today.[29] More notably, in 1950 there were no satellite data and fewer aircraft and ship-based reports. Thus, upper level weather conditions over the ocean were essentially unknown (a "true data void" as one meteorologist described it), making an earlier forecast impossible.[30]

SUPERSTORM 1950'S LONG-TERM IMPACTS ON STORM SURGE FORECASTING

Superstorm 1950 also improved a different type of model—those used to predict storm surge. Storm surge is a rise of seawater caused by the winds from a storm blowing water toward shore. It is most often associated with hurricanes, and it is their deadliest aspect. However, storm surge can be caused by strong midlatitude cyclones as well. Besides drowning people, it erodes beaches, floods coastal areas, and destroys structures.

In the 1960s, the National Weather Service began developing models to predict storm surge. These were done through the Techniques Development Laboratory (now the Meteorological Development Laboratory), which was formed in 1964.[31] Beginning in the 1970s, several models to predict storm surge from extratropical storms were developed.

Critical to development of these models were TDL researchers N. Arthur Pore and William S. Richardson. Pore and Richardson wrote numerous technical memorandums on storm surge modeling that relied on Superstorm 1950 as a test case.[32] These memorandums focused on the creation and improvement of storm surge models applicable to extratropical systems (or midlatitude cyclones) that affect the Northeastern coast. Often such storms are called nor'easters due to strong winds from the northeast.

One of the first models developed was described in a 1974 technical memorandum.[33] This seventy-page report described forecast equations that Pore, Richardson, and Herman P. Perrotti developed for ten locations from Portland, Maine, to Norfolk, Va., to predict surge associated with extratropical systems. Solutions to such equations could be done by a computer or calculated manually. To develop their model, the authors used data from sixty-eight storms from 1956 through 1969.

To test their model, they used five additional storms, including Superstorm 1950. This storm was of particular value, and its unique nature presented multiple challenges. The winds that created the storm surge were atypical, from the east and southeast instead of the northeast. The timing of the storm, coincident with a spring tide (associated with a full moon), made the tide higher than normal. Additionally, the gigantic size and intensity of the storm caused record high tides at more than twenty locations from Long Island to Maryland. Overall, the model underforecasted the peak surge in most cities (and had some timing errors), but in their words the calculations were "not too bad" given the storm's record-setting conditions and atypical wind field.[34]

Storm surge prediction efforts were further refined in the 1980s, and the 1950 superstorm was a critical part of that effort. A paper from 1980 concerning the development of a revised prediction equation for Charleston, S.C., once again used Superstorm 1950 as a test case.[35] Curiously, the storm caused a 3-foot *negative* surge in Charleston due to winds blowing offshore. The revised equation was able to predict this more accurately than the previous iteration. Three years later, two additional papers drew on Superstorm 1950 as a test case.[36] Unsurprisingly, the forecasts improved over the previous ones, though forecasting the storm's record high tides was still a challenge. By this time, computer models had become indispensable tools in surge forecasting, and there were numerous newer case studies available. However, the exceptional and unusual nature of Superstorm 1950 gave it lasting value as a test case.

THE 1950S WAS A TIME OF GREAT ADVANCES IN METEOROLOGY, WITH THE development of new understandings of the atmosphere and the first computer models. The weather disruptions caused by Superstorm 1950 were severe, but generally short-lived. The storm, however, lived on within meteorology. Its fortuitous timing meant that it played a large role in developing and perfecting meteorologists' and computers' understanding of the atmosphere. It was also important in extending knowledge of storm surge forecasting. Ultimately, while Superstorm 1950 killed hundreds and caused hundreds of millions of dollars in damage, countless lives and dollars have been saved since then due to the improved accuracy of weather forecasts. The quality of these forecasts is directly attributable to this storm.

10

NOW AND BEYOND

"While the effects of Sandy were devastating, FEMA [the U.S. Federal Emergency Management Agency] recognizes that it must plan for even larger, more severe storms and disasters."

— HURRICANE SANDY FEMA AFTER-
ACTION REPORT, JULY 1, 2013[1]

ON OCTOBER 22, 2012, THE EIGHTEENTH NAMED STORM OF THE HIGHLY active 2012 Atlantic hurricane season formed in the southwestern Caribbean Sea. Named Sandy, it strengthened to a Category 2 hurricane and moved northward, passing over Jamaica, Cuba, and the Bahamas, killing more than one hundred people, displacing hundreds of thousands of residents (especially in Haiti), and causing hundreds of millions of dollars in damage. Hurricane Sandy then spared the southeastern United States as it turned northeast, before merging with a cold front and turning back to the northwest. After merging with the front, the storm transformed into an exceptionally large and intense midlatitude cyclone. Tropical storm–force winds of 39 mph or greater stretched an incredible 580 miles from the center.[2] On October 29, Sandy finally made landfall in southern New Jersey before fading over southern Pennsylvania the next day. Some notable impacts of the storm included record tides approaching 14 feet, which caused extensive coastal flooding in New York and New Jersey, tropical storm-force winds from Maine to South Carolina, and several feet of snow with blizzard conditions in the eastern Appalachians.[3]

Sandy's transition from hurricane to midlatitude (or "post-tropical" as the National Weather Service called it) cyclone caused significant challenges for meteorologists and emergency management. Hurricanes are systems that originate in the tropics; they don't have fronts since all of the air is warm. However, when Sandy transitioned to a midlatitude cyclone, it became a weather system driven by contrasts between cold and warm air, with fronts, much like any other midlatitude cyclone. This caused numerous problems for the government agencies that warned of and responded to the storm.

Because it was no longer tropical in nature, forecasting and warning operations transitioned within the National Weather Service (NWS) from the National Hurricane Center to the Hydrometeorological Prediction (now Weather Prediction) Center, a different office more than 1,000 miles away by car. In addition, since it was no longer a hurricane, it no longer had a name (midlatitude cyclones aren't named). To help, the NWS referred to it as "post-tropical storm Sandy," a technically correct title but with a term largely unknown to non-meteorologists. FEMA still called it "Hurricane Sandy," a name that was both technically incorrect and one unable to capture the varied impacts of the storm, such as record-setting snow. The whole purpose of naming storms is to provide a simple method to reduce confusion.

Names aside, the evolution of the storm and switching of forecast centers made it a challenge to communicate the severity and intensity of Sandy to the general public. The general public knows that hurricanes are serious storms, but Sandy was no longer a hurricane. Furthermore, one hazard of the storm, heavy snow, is not something associated with hurricanes at all. Yet it was quite a serious storm, and the government and the general public still needed to make preparations.

Describing Sandy as a *superstorm* may have been the best way to alert the general public to its seriousness.[4] Although it originated in a vastly different environment than Superstorm 1950, by the time it made landfall in the U.S., Sandy was a midlatitude cyclone with record-setting disparate hazards, just like the 1950 storm. But what makes a superstorm a superstorm and not just a routine midlatitude cyclone? How does thinking about superstorms help us understand and prepare for future storms?

WHAT SETS A SUPERSTORM APART?

Superstorm 1950 began on Thanksgiving, or November 23, 1950, and ended early the following week. The storm was most severe on Saturday, November 25, when it buried much of Ohio, West Virginia, and western Pennsylvania in record-setting snow; encased west central Pennsylvania in ice; flooded central Pennsylvania, interior New England, and the New Jersey coast; blew down many trees and structures in the Northeast; and brought frigid conditions to the South. More than three hundred people died in connection with the storm. Like with modern disasters, differences in vulnerability meant that the impacts varied with race, gender, and social class. In this case, men, those living in poverty, and people of color suffered the greatest losses. At the time, it was the most expensive disaster on record.

Despite all of the negative impacts, Superstorm 1950 fortuitously occurred at the advent of computer modeling in meteorology. Its exceptional conditions inspired the modelers, who repeatedly used it as a case study to develop, test, and refine computer models over the next fifty years. Development of models has greatly improved forecast quality and saved countless lives.

Superstorm 1950 was an unusually strong midlatitude cyclone. Midlatitude cyclones are common regular occurrences in the United States (except tropical Hawaii). Most are short-lived with minor impacts.

They typically form between about thirty and sixty degrees north latitude—the north–south extent of the continental U.S. and adjacent Canadian provinces—and develop from sharp differences between cold polar and warm tropical air masses. Because these differences are greatest during winter, the most intense midlatitude cyclones in North America usually occur from October through March. This includes the four we will discuss here: Sandy (October 2012), Superstorm 1950 (November), the White Hurricane (November 1913), and the 1993 Storm of the Century (March).

Tropical cyclones are less common than midlatitude cyclones but are longer lived and more likely to have severe societal impacts. As the name implies, tropical cyclones, the strongest of which are hurricanes, form in the tropics. Since all tropical air is warm, these cyclones lack fronts and contrasting air masses. Instead, they are giant heat engines that transfer vast of amounts of energy from warm, tropical waters into the atmosphere. Since tropical ocean waters are hottest in the summer and early fall, this is the time of year when tropical cyclones are most numerous. Superstorm Sandy began as a tropical cyclone. Shortly before landfall, it transitioned into a midlatitude cyclone with fronts and other associated characteristics.

While the difference between midlatitude cyclones and hurricanes is clear, what distinguishes superstorms from other midlatitude cyclones? The distinguishing characteristic of a superstorm is its combination of large size, extreme weather, and societal impacts that affect a very large number of people. Superstorm 1950, Sandy, and the others discussed here all had multiple, record-setting disparate weather hazards. They also had significant impacts on large numbers of people. Table 6 summarizes the meteorological extremes and societal impacts of Superstorm 1950. Similar tables could be generated for other superstorms.

Meteorologically, Superstorm 1950 was one of the most intense November cyclones to strike the United States. It set snow, wind, and temperature records, and it is currently the top-ranked Midwestern

TABLE 6. Superlatives caused by Superstorm 1950.

Meteorological	Societal
All-time snowfall records in multiple cities.	7th deadliest nontropical storm–related disaster in U.S. history.
Regional Snowfall Index top-ranked snowstorm in the Midwest region and among top 10 for the Northeast region.	20th deadliest weather-related disaster in U.S. history (including tropical systems and heat waves).
All-time wind speed records set.	Only 2 storms since 1950, Hurricanes Audrey and Katrina, had a greater death toll.
All-time ice storm records set (unofficially).	Most costly disaster in U.S. history at the time of occurrence.
Numerous record lows and record cold highs (min maxes) set for Nov.	Affected 26 states.
Exceptionally cold temperatures aloft.[*]	Affected two-thirds of the U.S. population.
Record high tides along the New York and New Jersey coasts.[†]	Routine life disrupted for days to a week in most places.

[*]Grumm (2010a) and Hart and Grumm (2001).
[†]Harris and Lindsay (1957), as cited in Pore, Richardson, and Perrotti (1974).

snowstorm and within the ten most intense Northeastern snowstorms in the Regional Snowfall Index for each region.[5] Socially, Superstorm 1950 was among the deadliest and costliest storms to affect the U.S. and disrupted life for more than two-thirds of the country's population.

TWO ADDITIONAL CASE STUDIES CAN HELP DEFINE SUPERSTORMS: THE 1913 White Hurricane, a storm with many similarities to the 1950 storm, and the 1993 Storm of the Century, which was arguably the second biggest weather event of the twentieth century after Superstorm 1950.

The 1913 White Hurricane was among the most severe storms ever to strike the Great Lakes region. It is best known for its effects on shipping: twelve ships sank, thirty-one ran aground, and more than 250 sailors lost their lives.[6] Like Superstorm 1950, it occurred in November. Its path and intensity were similar to that of Superstorm 1950, but the blocking high pressure center did not materialize and the effects were more severe to the west of the storm's center. All of the Great Lakes were affected, but the strongest winds were over Lake Huron, where mariners estimated 90 mph wind gusts caused 35-foot waves and extensive coastal damage.[7] Meanwhile, more than 22 inches of snow buried Cleveland, with 17.4 inches of that falling in just 24 hours, setting a record for the time.[8] The city's electric and telephone systems were devastated to such an extent that fallen poles and wires caused more transportation disruption than the snow itself.[9] Much like with Superstorm 1950, heavy snow blanketed eastern Ohio, western Pennsylvania, and West Virginia, where more than 2 feet accumulated in Parkersburg.[10] Severe wind gusts were reported from Buffalo, N.Y., to Duluth, Minn. Meanwhile, record cold air pushed into the Southeast, where daily record lows in places like Huntsville, Ala., Atlanta, Ga., and Greenville, S.C., still stand more than a century later. (None of these were monthly records, and these records are not dramatically different from record lows set on other days in mid-November, unlike the highly anomalous records from 1950.)

The 1993 Storm of the Century was a massive superstorm that affected the entire Eastern Seaboard and most of the Gulf Coast. More than two hundred people died, and it caused $2 billion (1993 dollars) in damages.[11] The areal extent of heavy snow was unmatched in the twentieth century, and the snow blanketed highly populated areas, causing great disruption; the storm ranks No. 1 on the Northeast Snowfall Impact Scale.[12] New snowfall records were set from Upstate New York, where nearly 43 inches buried Syracuse, to Birmingham, Ala., where the greatest snowfall ever recorded fell (13.0 inches).[13] Besides the snow, record low temperatures were observed from northern Maine to the Gulf Coast, and hurricane-force wind gusts were observed from Massachusetts to Louisiana. In Florida, a derecho (a long track windstorm associated with a line of thunderstorms) brought hurricane-force winds to the entire peninsula before crossing the Straits of Florida and ravaging Cuba, where winds gusted over 100 mph and more than $1 billion in damage occurred.[14] If the Storm of the Century wasn't a superstorm, perhaps nothing is.

DEFINING SUPERSTORMS

Using Superstorm 1950 and the other three storms as examples, we can now establish four characteristics that distinguish superstorms from ordinary storms.

1. SUPERSTORMS ARE MIDLATITUDE CYCLONES.

Midlatitude cyclones are caused by differences between cold and warm air and have fronts associated with them. In contrast, hurricanes are cyclones that originate from the tropics and do not have fronts, though they may seed midlatitude cyclones with moisture and energy, as the October 1991 Perfect Storm did, or evolve, as Sandy did. This also distinguishes superstorms from smaller events such as thunderstorm complexes, lake-effect snow events, and downslope windstorms.

2. SUPERSTORMS CAUSE SIGNIFICANT IMPACTS OVER A LARGE AREA.

The term *large* is somewhat vague; at a minimum we can think of large as being 1 million square kilometers, or 386,000 square miles, or approximately twice the size of California. The best way to measure the size of a storm is subject to debate, so an easier way to assess size is through the use of National Oceanic and Atmospheric Administration (NOAA) climate regions (see fig. 10). Although climate regions vary in size, most storms that set records in three regions are likely to meet or exceed the size criterion. This isn't a perfect explanation; a small but intense storm that occurs where regions come together could meet this

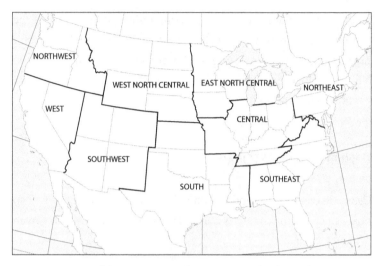

FIGURE 10. Map of the nine NOAA climate regions. (Redrawn from Karl and Koss [1984].)

criterion. But use of climate regions as a proxy for size is a simple and practical way to filter out most small storms.

All four examples meet or exceed this criterion. The White Hurricane of 1913 set records within three regions (East North Central, Central, and Northeast). Superstorm 1950 set records in five different climate regions of the eastern and central United States. The 1993 Storm of the Century affected the Northeast, Southeast, and Central U.S. Sandy also affected those three regions—Northeast, Southeast, and Central.

An exception to this criterion pertains to coastal storms that affect only the Northeast and Southeast climate regions. Nearly 40 percent of the U.S. population lives in the Northeast and Southeast regions, so a storm affecting the entirety of these regions would have an outsize impact on the country as a whole. Sandy illustrates the importance of this exception—had Sandy's snow fallen in Pennsylvania instead of West Virginia, it would have affected only two regions.

3. SUPERSTORMS FEATURE MULTIPLE AND DISPARATE RECORD-SETTING HAZARDS.

All four example storms had a heavy snow component and also included a combination of several of these: damaging winds, record cold, coastal damage/flooding, and severe convection. While it is not surprising that snow and record cold may happen simultaneously with a storm, damaging winds, flooding, and severe convection occur under very different processes, hence the use of the term *disparate*. (Conversely, hazards such as damaging wind, hail, and tornadoes often occur in connection with each other.) Furthermore, to be a superstorm, each hazard should be, in and of itself, record-setting. Each of these storms set records for multiple hazards.

4. SUPERSTORMS CAUSE SIGNIFICANT DISRUPTION TO SOCIETY IN TERMS OF PROPERTY DESTRUCTION, LIVES LOST, OR ECONOMIC DISRUPTION.

The White Hurricane and Superstorm 1950 were quite deadly; Superstorm 1950 also caused destruction in much of the eastern United States. The 1993 Storm of the Century paralyzed the heavily populated East Coast and Northeast Corridor, disrupting flights and commerce nationwide. Sandy was the most expensive disaster at the time it occurred and affected many of the same areas as the 1993 storm.

We can now combine these four criteria into a formal definition of *superstorm*:

> A superstorm is a midlatitude cyclone that causes record conditions in multiple disparate meteorological hazards over a large area. Superstorms also cause significant societal impacts via loss of life, destruction of property, or general disruption.

Each of the four storms discussed meets the definition of a superstorm. However, this definition should be viewed as a starting point; I am calling on the meteorological community to refine it. Determining more detailed meanings for terms such as *large* and *significant* would be a logical place to begin.

THE SAME STORM TODAY

Why should we care about a storm that occurred more than seventy years ago, in a society that was quite different from that of today? If Superstorm 1950 were to occur now, how would the impacts be different?

Great progress has been made in racial relations, the reduction of poverty, and gender equality. Nonetheless, race, social class, and gender continue to determine who suffers the most in disasters. And despite great advances in snow removal and technology in general, society at large remains highly vulnerable to several hazards associated with the storm, especially ice accumulation and wind. In addition, modern disasters have shown that the cost of property damage is much greater today.

On the plus side, improvements in forecasting would greatly reduce the disruption from all storm-related hazards. Even the 1993 superstorm was well-forecast with "unprecedented lead times," allowing for preparations by government and citizens.[15] Forecasting, and the general public's confidence in forecasts, has continued to improve since then. This confidence means that if a superstorm were forecast today, schools, businesses, and other organizations would cancel events or close entirely, helping to reduce traffic volume and making it much easier for authorities to clear roads. Such an effect has already been observed with modern snowstorms, such as the Boston Blizzard of 2013.[16] In 1950, many people became stuck attempting to travel to and from college football games, but today major sporting events are more likely to be canceled. For example, from 2000 through 2019, the NFL postponed nine games due to hurricanes, snow, and other weather-related events. Some of these were postponed simply in anticipation of inclement weather. With better forecasting, people today also take action to reduce their vulnerability. Stores are overrun as citizens rush to buy supplies like bread and toilet paper, but this has a net positive effect of reducing the need to travel during the worst of the storm. People in flood-prone areas or along the coast can move to higher ground or evacuate. Hunters might go home and not be trapped by floodwaters, as was the case for many hunters in New Jersey in 1950. Those in areas expected to get high winds can take preventative actions like securing loose objects and boarding up windows.

However, although we have made significant advances in forecasting that would greatly reduce the total disruption from a storm like Superstorm 1950, specific hazards would still cause problems, sometimes significant.

CHANGES IN SNOW AND ICE IMPACTS

If a similar storm occurred today, technological advances would greatly reduce the travel disruptions from excessive snow. First, modern automobiles and, more importantly, modern snow-clearing equipment are much less likely to suffer from stalling or other mechanical failures. Second, tires have greatly improved, allowing for better traction on snow-covered roadways. Third, many people heed forecasts and choose to stay home during extreme weather. These changes combine to create a positive feedback loop. Fewer stalled or disabled vehicles mean that road crews have an easier time clearing the roads. With less snow on the roads, vehicles are less likely to get stuck in the snow (plus the tires provide better traction), allowing for additional snow-clearing gains. In addition, with fewer vehicles on the road, it is easier to clear the roads. Also, the quality and quantity of municipal equipment available to clear snow has increased. Of course, with amounts of snow like those in 1950, people could not leave their homes or other locations immediately; road crews need time to clear record-setting amounts of snow. Nonetheless, people would be stuck in place for a much shorter period.

A similar ice storm would still cause major problems today. The root cause is that our society remains highly vulnerable to a loss of electricity, and, if anything, societal vulnerability has increased since 1950. An amount of freezing rain similar to that in 1950, which was much greater than a typical ice storm, could be expected to destroy high-voltage transmission lines. In 1998, an ice storm with similar accumulation amounts did destroy high-voltage transmission lines in Quebec. Nearly 1.4 million customers lost power there, and it was Canada's most expensive natural disaster at the time. Some customers did not have power restored for nearly a month. The storm also had significant effects in eastern Ontario, New Brunswick, and northern New England. In Maine, 80 percent of the population lost power, and some outages persisted for more than two weeks.[17]

Another concern from ice is the risk of a long-lasting regional blackout. Electric industry observers have repeatedly expressed concerns about the overall hardiness of the grid for years, and in 2021 an outbreak of cold air in Texas nearly caused a complete collapse of the power grid in that state. Multiple media outlets noted that the grid was within 5 minutes of this disaster. And, according to the grid's operator, the Electric Reliability Council of Texas, a complete collapse could have caused a blackout lasting weeks or even months. Determining the risk of a complete collapse of the electric grid is beyond the scope of this book, but a severe ice storm and record cold outbreak like that from Superstorm 1950 might trigger such a doomsday scenario.

On an individual level, a blackout today is more difficult to cope with.[18] People who lost power in 1950 were remarkably innovative in how they adapted motors from other products to generate power and operate equipment like gas pumps and milking machines. Today's motors and machines are less easily modified in this manner. Another concern is that most people without power today have no heat source. In 1950, most residents of Altoona had coal furnaces or other heating systems that did not require electricity. Today, nearly all heating systems, even if they use natural gas or propane, require electricity to operate. In addition, more than 25 percent of U.S. homes are all-electric, meaning that there is no alternative way to create heat if there is no electricity available. Furthermore, the percentage of all-electric homes is increasing.[19] Without heat, many people would have to evacuate their residence and stay in a shelter. Others might travel to areas with power and stay in a hotel or with family or friends. Such actions would increase the number of vehicles on the road during a hazardous event. With people staying in hotels, utility companies will experience difficulty in securing close lodging for their temporary workers.

Local or regional gasoline shortages are also likely because of reduced supply and increased demand. A superstorm's large size means that distant refineries or pipelines may be disrupted by other storm impacts, reducing supply. Locally, trucks may have difficulty making deliveries due to downed trees, snow, and/or ice. Without electricity, few gas stations can operate their pumps. Demand may also increase. People traveling from areas without power or buying fuel for generators increase supply pressures, too.

While generators may seem like a solution to power outages, homes with a generator still represent a small percentage of residences. Furthermore, the high cost of purchasing, installing, and maintaining a generator renders them unaffordable for most homeowners. Generators are simply not an option for renters.

CHANGES IN FLOODING IMPACTS

The riverine flooding associated with a modern storm should be less severe than that in 1950. Lock Haven, Pa., where in 1950 3 feet of water flooded more than 75 percent of the city, now has a levee, as do South Williamsport and Tyrone. Even Renovo, Pa., with a population around twelve hundred, now has some flood protection thanks to an upstream dam. Assuming that none of these structures fails, flooding in Pennsylvania is likely to be isolated in nature. A similar reduction in flooding is probable for interior New England as well.

In contrast, coastal flooding today would be worse than in 1950, and many times more costly. Coastal areas have seen explosive growth in population and property values, greatly increasing society's vulnerability.

New Jersey's population from 1950 to 2020 nearly doubled, from 4.8 million to 9.3 million residents, and neighboring Delaware's population has more than tripled. Moreover, the significant rise in property values during that time, especially for places near the shoreline, has greatly increased the losses from modern hurricanes and midlatitude cyclones.[20] Superstorm Sandy vividly illustrates this increase. Sandy caused $65 billion in damage (2012 dollars), primarily in coastal areas, a cost that dwarfs Superstorm 1950, even when the costs are adjusted for inflation.

Several larger environmental changes contribute to making modern coastal floods more severe. Climate change is causing sea level to rise; sea level has risen by around 6 inches since 1950.[21] Loss of coastal wetlands also increases the risk of flooding. An average of 80,000 acres were lost annually from 2004 to 2009, a 33 percent increase from the previous rate of loss from 1998 to 2004.[22] With these environmental changes, a similar storm will definitely cause more coastal flooding. When coupled with the increase in vulnerability due to coastal development, the economic costs of such a storm would be staggering. On a positive note, the number of deaths is likely to be smaller. For example, fewer New Jersey residents died from Sandy (ten) than from Superstorm 1950 (dozens), despite the increase in population from 1950 to 2012.

CHANGES IN WIND AND COLD IMPACTS
Wind damage from a similar contemporary storm would likely be comparable in magnitude and destruction. Again, downed trees will disrupt traffic and cause widespread power outages, and buildings and other structures will sustain damage. People without electricity will need to travel to shelters or other locations. Like with coastal flooding, the increase in property values means that costs will be much greater. Recent research has found that average annual windstorm losses have increased significantly in recent years.[23]

Record cold would be a relatively minor hazard for most citizens. Carbon monoxide poisoning might kill some people, but carbon monoxide deaths have been falling for decades due to public education programs and advances in motor vehicles.[24] Modern homes are better insulated and more airtight than older homes, so fewer water pipes will freeze. Antifreeze is cheap and plentiful, reducing the loss to vehicles, tractors, and other equipment. In contrast, farmers would still suffer severe crop losses, and the losses may be greater. Crop yields per acre have increased since 1950, causing greater economic risk when the weather is unfavorable. Additionally, a warming climate has decreased the average number of days below freezing, meaning that some farmers are less prepared for a record-setting early season cold outbreak.[25] Although the cold itself might not be a big deal for individuals, it could cause power problems and rolling blackouts. Much of the power equipment in Texas, for example, is not sufficiently hardened, or protected, from extreme cold, and this was a significant reason the electric grid nearly collapsed in the February 2021 cold outbreak. Many power generating stations were inoperable, yet demand was much greater than normal.

NO CHANGES IN RACIAL, CLASS-BASED, AND GENDER IMPACTS
Finally, the racial, class-based, and gender impacts of the storm would likely be similar. People of color and those whose income is below the federal poverty threshold disproportionately continue to experience longer and more severe impacts from disasters. While poverty rates for African Americans have decreased substantially since 1950, they remain more than double those of non-Hispanic whites (19.5 vs. 8.2 percent) and greater than all other major ethnic groups. Likewise, the median income for African Americans is only about 60 percent that of non-Hispanic whites. More generally, people whose income is below the federal poverty threshold are more likely to live in substandard housing with less insulation and more air leakage. Heating equipment may be inefficient, poorly maintained, and less reliable. In addition, they are more likely to suffer direct economic harm from missing work. An hourly worker who is unable to work because of inclement weather may lose a day's wages or more, while a person with a salaried position may not suffer any economic harm. People with fewer resources have fewer choices available should they need to leave their home and

shelter elsewhere. And during power outages, they are more likely to rely on dangerous heating methods such as a gas stove or portable propane space heater; a generator is unlikely to be an option. These are just some of the ways that a future superstorm would affect people of color and people whose income is below the federal poverty threshold more severely.

A modern version of Superstorm 1950 can be expected to kill men at a higher rate than women. First, the majority of disaster victims in the U.S., regardless of disaster type, are men. We need look no further than Superstorm Sandy, which occurred in 2012, to see that men made up 61 percent of deaths, almost double the rate for women (34 percent; gender was unknown for 5 percent of deaths).[26] In addition, the leading cause of death in Superstorm 1950, snow shoveling, remains much deadlier for men than for women. Men suffer cardiac-related injuries at three times the rate of women when shoveling snow, and men account for nearly all cardiac-related injuries associated with the use of snowblowers.[27] In reviewing the other forms of death from 1950, men today continue to be more likely to die in traffic crashes or drown in floods when compared to women. The next superstorm will be more deadly for men than for women.

WE HAVE SEEN THAT SUPERSTORM 1950 WAS NOT JUST AN ORDINARY MID-latitude cyclone, but an extraordinary one. Most of the meteorological records it set have not been surpassed in the more than seventy years since, and it remains one of the most widespread, disruptive, and deadly storms to affect the eastern United States. And although superstorms are rare, Superstorm Sandy shows that they remain highly disruptive, deadly, and immensely expensive, and these effects continue to vary with race, social class, and gender.

Climate change is causing an increase in extreme weather, and as a result our vulnerability to disasters is increasing. Learning more about superstorms will help us better prepare for the next one. It is only a matter of time. We need to be ready.

ACKNOWLEDGMENTS

WHILE WRITING ITSELF IS OFTEN AN INDIVIDUAL PURSUIT, WRITING A BOOK requires the contributions of many people behind the scenes.

I first conceived of this book while working on my PhD, but my advisor, Mark Monmonier, a seasoned book author himself, wisely redirected me to a smaller dissertation project. I thank him for his help years later with my proposal and other author-to-author advice.

I did the bulk of the research during a sabbatical leave, and I thank Ball State University for granting me that time. Ted Baker and the staff at the Innovation Connector were gracious hosts during my sabbatical; working alongside a group of nonacademics helped broaden my arguments and make them more relatable. My department chairs, Kevin Turcotte and Petra Zimmermann, have been enthusiastic supporters of this project, and I'd like to thank them and my colleagues for listening as I refined my arguments and worked through writing challenges. Thanks also to Mike Hradesky, geospatial specialist at Ball State, for making the maps, and to graduate student Sara Marzola Lippi for collecting information.

Good librarians are essential. Most critical for this effort were the Interlibrary Loan personnel at Ball State, including Elaine Nelson, supervisor, and Jessica Hatton, Jodi Sanders, Karin Kwiatkowski, Kerri McClellan, Kyla Hedge, and Lisa Johnson. They worked tirelessly to locate more than sixty items, many of which were small-circulation newspapers and rare books. Librarians and archivists at the Ohio History Center, West Virginia University, and the Heinz History Center were great resources, as were employees at public libraries throughout the Ohio Valley. Josh McConnell at the Altoona Public Library was particularly helpful in obtaining information about the ice storm.

I finished the manuscript in March 2020, just in time for COVID-19 to upend the publishing industry. This made it difficult to find a publisher, but I knew I had a good partner when Justin Race enthusiastically replied to my initial inquiry. He, Andrea Gapsch, Kelley Kimm, and all of the staff at Purdue University Press have been outstanding. Thanks for answering my many questions and calming me during my anxious moments.

Finally, thanks to my family for their love and support. My father, Tom Call, proofread the entire volume in a little more than month, catching numerous writing errors. And my wife and children gave me countless hours to work quietly in the "executive suite," or, as it was known prior to COVID-19, the master bathroom. Thank you.

NOTES

Introduction

1. For more on the FDRA see Sylves (2020, chapter 3). For more on the creation of FEMA, see books by Kneeland (2020 and 2021).

Chapter 1

1. Brunkard, Namulanda, and Ratard (2008). These researchers did not examine poverty specifically, but it also played a major role in death rates.
2. See Cutter, Boruff, and Shirley (2003) for a discussion of seventeen factors that increase or decrease a person's vulnerability. For a broader view of the relationship between hazards, vulnerability, and environmental justice, see the eponymous book by Cutter (2006).
3. The beating took place in 1946 and the trial in 1947. The trial was considered a sham and called a "travesty" by none other than the presiding judge, J. Watries Waring, who will return as a minor character in chapter 8. In a sign of how much race relations have improved since 1946, a historical marker was dedicated in 2019 at the site where the beating occurred. The ceremony was attended by numerous governmental officials, including the local representative to Congress. The Batesburg-Leesville mayor also publicly apologized for the mistreatment of Woodard.

Chapter 2

1. Description of Superstorm 1950, as quoted in Schwartz (2007).
2. Based on Changnon et al. (2008).
3. Kocin and Uccellini (2004, 346).
4. Bristor (1951).

5. Bristor (1951).

6. Schmidlin and Schmidlin (1996).

7. Value based on Kocin and Uccellini (2004), and North American Reanalysis data. Bristor (1951) suggested wind speeds around 200 mph in his analysis.

8. Smith (1950).

9. For the clearest view of the paths of the Labrador high and storm itself, see charts 2 and 3 in Aubert (1950).

10. Grumm (2010a) discusses the role of the atmospheric river in more detail.

11. For more on bomb cyclones, see Sanders and Gyakum (1980).

12. Smith (1950).

13. Bristor (1951) describes this best; figure 1b is particularly beneficial.

14. Smith (1950).

15. Dayton pressure record reported by Smith (1950); inches of mercury reading from Schmidlin and Schmidlin (1996).

16. For more on the Regional Snowfall Index, see Squires et al. (2014). For rankings of storms, visit the RSI web page of the National Oceanic and Atmospheric Administration's (NOAA) National Center for Environmental Information at https://www.ncdc.noaa.gov/snow-and-ice/rsi/.

17. Totals from contemporary climatological reports. The climatological report from West Virginia lists an incorrect snow total for Pickens in the opening narrative, but the correct amount is listed in the data table later in the report.

18. Grumm (2010a, 2010b).

19. Grumm (2010a).

20. James (1952); Smith and Roe (1952).

21. Bristor (1951); Endlich (1953).

22. Grumm (2010a).

23. Hart and Grumm (2001).

24. For more on this storm, see Brown (2002).

25. Walz (1913); Mook (1949).

26. Mook (1949).

27. For more on predictability of the storm see Phillips (2001, 2004). Kistler et al. (2001) describe Superstorm 1950 as "quite predictable."

Chapter 3

1. Pickenpaugh (2001, 9).
2. Recorded both by Pickenpaugh (2001, 62) and the *Cleveland Plain Dealer*.
3. Pickenpaugh (2001, 7) and the *Pittsburgh Press*.
4. Reported by Pickenpaugh (2001, 12) and the *Cleveland Plain Dealer*.
5. Pickenpaugh (2001, 13–15).
6. People having been complaining about Christmas creep for generations, but most evidence is anecdotal outside of Kelly (2008). However, the Christmas shopping season did creep earlier in 1939 when President Franklin D. Roosevelt unilaterally changed the date of Thanksgiving from the last Thursday to the next-to-last Thursday of November, making it November 23 that year. (After a few years of confusion and multiple "Thanksgivings," in 1941 Congress proclaimed it to be to the fourth Thursday of November.) Thus, the year 1950 was only the second time since the official change that Thanksgiving was not the last Thursday of the month.
7. This story was found in the *Cleveland Plain Dealer* and the *Pittsburgh Press* and recounted in Pickenpaugh (2001, 7), but the accounts disagree on the value of the generators, and whether or not the man represented himself as a driver or was simply sleeping in the cab. Only the *Plain Dealer* reported on the later finding of the truck itself.
8. *Columbus Dispatch* and Pickenpaugh (2001, 45).
9. Pickenpaugh (2001, 51) has a fuller discussion of this.
10. Pickenpaugh (2001, 50). The NCAA record for punts in a game is thirty-nine by Texas Tech. This occurred in a 0–0 game vs. Centenary in Shreveport, La. The game, on November 11, 1939, was played in torrential rain. Centenary punted thirty-eight times.
11. Pickenpaugh (2001, 45).
12. Pickenpaugh (2001, 45). A similar argument was made by the *Columbus Dispatch*'s football columnist, Russ Needham.
13. As reported by the appropriately named *Oil City (Pa.) Blizzard*.

Chapter 4

1. Quote from an "emergency message broadcast last night by Mayor David L.

Lawrence," reprinted in the *Pittsburgh Press*, November 27, 1950, p. 12.

2. For a more on these trends, see Kneeland (2021, chapters 1 and 2).

3. Two examples where a snowstorm ruined a politician's future were in New York City (John Lindsay in 1969) and Buffalo (Stan Markowski in 1977). (Both are elaborated on in Kneeland 2021, 62–64.) Disasters are not always fatal to mayors, however (for examples, see Call 2004, 51; and Bodet, Thomas, and Tessier 2013). Nonetheless, taking responsibility when disaster strikes is a good political strategy (Miller and Reeves 2021).

4. For more on this, see Call (2005).

5. The deaths were as follows: a 4-year-old boy near Pittsburgh (as reported by the *Pittsburgh Press*); a 67-year-old man and his 20-month-old granddaughter at a rural farmhouse near Butler, Pa. (as told by the *Butler Eagle*); two children in a house fire in West Virginia (listed in the *Charleston [W.Va.] Gazette*); and two children in Kentucky (described by the *Louisville Courier Journal*). In all cases, faulty heating equipment caused the fires. Poor road conditions contributed to the three deaths in western Pennsylvania.

6. The final number of deaths in Ohio was estimated at "up to seventy" by Schmidlin and Schmidlin (1996).

7. Haney (2017).

8. Pennsylvania and Delaware are the only states that currently use the term *prothonotary* to describe the clerk of court, something that causes confusion among even the highest government officials. In an anecdote from the *Post-Gazette*, when visiting Pittsburgh in 1948, President Harry S. Truman was introduced to the local prothonotary, David B. Roberts, and responded by saying, "What's a prothonotary?" Over the years Truman's question has evolved into "What (in/the) hell is a prothonotary?!" a humorous way of expressing most people's ignorance about the term.

9. For evidence of how much less safe cars were, see Insurance Institute for Highway Safety (2009).

10. Cecil Kreps, ASE Certified Mechanic at Earl's Auto Repair, Muncie, personal comment (December 15, 2017).

11. As noted by local papers in all aforementioned cities except for Steubenville (Schmidlin and Schmidlin 1996).

12. The story was recounted in both the *Columbus Dispatch* and Pickenpaugh (2001, 53), but they disagree on the whether there were one or two arrests.

13. Age information from Zeller (1998), who also characterized the original building as a "navy surplus recreation center."

14. Some media accounts erroneously listed his age as 27, but his birthdate was March 19, 1924, according to a search of the Social Security Death Index. Facts for this paragraph were compiled from the *Cleveland Plain Dealer* and some of the newspapers that ran syndicated stories on Andras, including the *Cincinnati Enquirer, Circleville (Ohio) Herald, Indianapolis News,* and *Newark (Ohio) Advocate.*

15. Death information was obtained from the Social Security Death Index and an obituary published in the *Mansfield (Ohio) News-Journal* on September 30, 2003.

16. Numerous battles between the Ohio governor and other politicians over financing, routes, and other issues delayed construction of the Ohio Turnpike, which had not even begun in 1950. As the *Cleveland Plain Dealer* described it in an editorial on December 1, 1950, "While Pennsylvania keeps lengthening its turnpike, Ohio keeps lengthening the delay in starting one for the state."

17. Motorists searching for parking in congested downtown areas is a problem in many cities. A summary of sixteen studies found that, on average, 30 percent of cars in these areas are searching for parking (Shoup 2005, 290). With parking bans and piles of snow further limiting the amount of parking available, the percentage of cars searching for parking was surely greater than average.

18. These traffic jams presaged how congestion from automobiles, the convenience they provided compared to mass transportation, and the lack of easily available parking contributed to the eventual decay of downtowns as centers of commerce and retail. For a case study of this process, see Tolle (2012).

Chapter 5

1. *Altoona Tribune*, December 7, 1950.

2. The descriptor "west central Pennsylvania" for Clearfield and Blair Coun-

ties is simply used for convenience. It is a not a real region. These counties are not connected in any cultural or physical sense; they barely touch each other.

3. For more on this see Call (2010).
4. *Climatological Data, Pennsylvania, November 1950*, U.S. Department of Commerce, Weather Bureau.
5. In something that resembles a movie plot, D. W. Jardine had reached the "normal retirement age" of 65 and was originally scheduled to retire on November 1, 1950. In September 1950, the board of directors asked him to remain one more year to oversee construction of several new power plants, according to the *Clearfield Progress*. The *Altoona Tribune* reported that he finally retired in late 1951.
6. U.S. Bureau of the Census (1953), Vol. 2 (Pa.), table 20.

Chapter 6

1. Perry (2000).
2. The March 1936 record high flood crest of 29.23 feet has been exceeded only once. Hurricane Agnes, one of Pennsylvania's most expensive natural disasters, caused a crest of 33.27 feet in 1972.
3. Pore, Richardson, and Perrotti (1974).
4. The African American population was described by the Works Progress Administration (1939, 639).

Chapter 7

1. As reported in the *Burlington Free Press*, November 30, 1950.
2. In 1950 it was referred to as the North Bridge, but I use its later name to avoid confusion with the 1973 George N. Wade Memorial Bridge, which locals refer to as the North Bridge. The M. Harvey Taylor Bridge opened in early 1952.
3. For more information on the slaying, see Jenkin (1978) and Rattigan (2005).
4. Delaney (2000).
5. Hunt (1962).
6. Delaney (2000); Nagashybayeva (2009).
7. For more on the early history of FEMA, see Kneeland (2020, 2021).

Chapter 8

1. Quote from *Climatological Data, Alabama, November 1950.*
2. Berko et al. (2014). Measuring excess deaths caused by extreme heat or cold is difficult; see Dixon et al. (2005) for an overview of the challenges.
3. It is inaccurate to reference "wind chill" here as wind chill temperatures are simply an attempt to quantify the cooling effects of wind as experienced by humans. They are not real temperatures felt by objects. No matter how strong the wind, if the actual air temperature is above 32°F, water will not freeze. (A strong wind will cool objects more quickly, though.)
4. U.S. Bureau of the Census (1953) Vol. 1, p. 1-26.
5. U.S. Bureau of the Census (1953) Vol. 1, p. 1-66.
6. A stoker is a device that automatically feeds coal to a stove or furnace. Informally, the term may be used to refer to a coal stove or furnace. A flue is the opening in a chimney through which exhaust gases are vented. A flue may simply be a metal pipe if there is no chimney.
7. U.S. Bureau of the Census (1953) Vol. 1, p. 1-26.
8. An interesting footnote is that despite the bitter cold, more than 125 people (mostly African Americans) gathered for the presentation of a citation to U.S. District Court Judge J. Watries Waring at his house on November 26. His rulings against school segregation and in favor of opening South Carolina Democratic primaries to all races caused him to be celebrated by Blacks and loathed by many whites. Widely shunned as a result, he left Charleston, S.C., in 1952. For more on Judge Waring, see Yarbrough (1987).
9. Rates from Morrill (2015).
10. Smith (1950) and weekly crop bulletins.
11. Information from weekly crop bulletins. Tangerines freeze at warmer temperatures and shorter time intervals than other citrus crops.
12. Information from weekly crop bulletins.
13. Lange, Olmstead, and Rhode (2009).
14. Lindop and De Cupua (2009, 47). Diners Club, the first general credit card, had just been created earlier in 1950; at that time, it could only be used at fourteen restaurants in New York.
15. Carbon monoxide poisoning has since become more common from the improper use of electric generators. It is surprising, given the substand

heating equipment often in use, that it was rare in 1950. Perhaps this is because houses with substandard heating systems were also unlikely to be airtight.

Chapter 9

1. Quote delivered at the Symposium on Modern Concepts in Meteorology at the National Academy of Sciences. Entire remarks are in Charney (1955).
2. Wind and damage report from a speech by Phillips (2004).
3. Cox (2002, 159).
4. Blum (2019, 34). The reason for the coordinated weather observation in 1910 was a mistaken belief that Halley's Comet would affect the weather.
5. Richardson (1922, 219). A more recent estimate suggested that 204,800 would be needed (Cox 2002). Richardson was also missing some important theoretical knowledge (Phillips 2001).
6. Richardson (1922, 219).
7. Cox (2002, 203).
8. Charney (1947).
9. See Charney (1949) for his views on NWP.
10. McCartney (2001, 101).
11. McCartney (2001, 94).
12. McCartney (2001, 94).
13. McCartney (2001, 101–2).
14. Lynch (2008).
15. Phillips (1951).
16. Charney (1955).
17. Gilchrist (2006).
18. Gilchrist (2006).
19. Gilchrist (2006).
20. Charney and Phillips (1953).
21. Charney (1954).
22. Phillips (1958).
23. Charney (1954).
24. Cox (2002, 162).

25. Phillips (2001).

26. For more on improvements in meteorology and models during the twentieth century, see Blum (2019). For an example of how these improvements in forecast quality have benefited society, see Call, Grove, and Kocin (2015). That paper compares the impacts of two meteorologically similar Boston blizzards, one in 1978 and the other in 2013. Due to improved forecasting, the 2013 blizzard caused a much shorter and less severe disruption.

27. For more on some of the critical players, see Cox (2002) or Fleming (2016).

28. Kistler et al. (2001).

29. Kistler et al. (2001).

30. See Kistler, Uccellini, and Kocin (2004). Norm Phillips himself also made a similar comment in a speech in 2004.

31. Meteorological Development Laboratory (2019).

32. See Pore, Richardson, and Perrotti (1974); Richardson and Boggio (1980); and Richardson and Gilman (1983a, 1983b).

33. Pore, Richardson, and Perrotti (1974). For a less technical summary of this report, see Pore and Barrientos (1976).

34. Pore, Richardson, and Perrotti (1974).

35. Richardson and Boggio (1980).

36. Richardson and Gilman (1983a, 1983b).

Chapter 10

1. Federal Emergency Management Agency (2013, 36).

2. Federal Emergency Management Agency (2013, 4).

3. For a visual representation of Sandy's disparate impacts, see Halverson and Rabenhorst (2013).

4. An article in *Weatherwise*, a magazine meant for weather enthusiasts, did call it a *superstorm* (Halverson and Rabenhorst 2013).

5. For more on the RSI, see Squires et al. (2014).

6. Brown (2002, xvii).

7. A modern model simulation by Wagenmaker and Mann (2013) found those estimates plausible.

8. Armington (1913).

9. Brown (2002, 162).

10. Walz (1913). Coincidentally, many places that received significant snow from this storm were the same ones hard hit by Superstorm 1950.

11. This number includes indirect deaths (from snow shoveling, for example), but these were not consistently reported; this number is merely an estimate. For more about the issues with counting deaths from this storm, see National Weather Service (1994).

12. Kocin and Uccellini (2004, 257).

13. Kocin et al. (1995).

14. Alfonso and Naranjo (1996).

15. Uccellini et al. (1995).

16. Call, Grove, and Kocin (2015).

17. National Weather Service (1998).

18. See Call (2007) for a more detailed discussion.

19. U.S. Energy Information Administration (2018).

20. Pielke, Jr., et al. (2008).

21. The website https://sealevelrise.org/ has an exact number for this; https://www.climate.gov/news-features/understanding-climate/climate-change-global-sea-level provides a government-determined value.

22. Dahl and Stedman (2013).

23. Changnon (2011).

24. Hampson (2016). Government mandates that require the installation of carbon monoxide detectors may also be helping.

25. Karl et al. (2009, 112).

26. Centers for Disease Control (2013).

27. Haney (2017). While the total number of injuries suffered from snow blowing is much less than that from shoveling, it is unclear if this is because snow blowers save lives or because relatively few households own snow blowers.

REFERENCES

Alfonso, A. P., and L. R. Naranjo. 1996. "The 13 March 1993 Severe Squall Line over Western Cuba." *Weather and Forecasting* 11 (March): 89–102. https://www.spc.noaa.gov/misc/AbtDerechos/papers/Alfonso_1996.pdf.

Armington, J. H. 1913. "Climatological Data for November, 1913: District 4, the Lake Region." *Monthly Weather Review* 41, 1678–87.

Aubert, E. J. 1950. "The Weather and Circulation of November 1950." *Monthly Weather Review* 78 (11): 201–3. https://doi.org/10.1175/1520-0493(1950)078<0201:TWACON>2.0.CO;2.

Berko, J., D. D. Ingram, S. Saha, and J. D. Parker. 2014. *Deaths Attributed to Heat, Cold, and Other Weather Events in the United States, 2006–2010.* National Health Statistics Reports No. 76. https://stacks.cdc.gov/view/cdc/24418.

Bristor, C. L. 1951. "The Great Storm of November, 1950." *Weatherwise* 4 (1): 10–16.

Blum, A. 2019. *The Weather Machine: A Journey Inside the Forecast.* Ecco.

Bodet, M. A., M. Thomas, and C. Tessier. 2016. "Come Hell or High Water: An Investigation of the Effects of a Natural Disaster on a Local Election." *Electoral Studies* 43, 85–94. https://doi.org/10.1016/j.electstud.2016.06.003.

Brown, D. G. 2002. *White Hurricane: A Great November Gale.* McGraw-Hill.

Brunkard, J., G. Namulanda, and R. Ratard. 2008. "Hurricane Katrina Deaths, Louisiana, 2005." *Disaster Medicine and Public Health Preparedness* 2 (4): 215–23. https://doi.org/10.1097/DMP.0b013e31818aaf55.

Call, D. A. 2004. "An Integrated Hazard Analysis of Ice Storm Impacts, Warnings, and Emergency Plans." MA thesis, Syracuse University.

Call, D. A. 2005. "Rethinking Snowstorms as 'Snow Events': A Regional Case Study from Upstate New York." *Bulletin of the American Meteorological Society* 86 (12): 1783–93. https://doi.org/10.1175/BAMS-86-12-1783.

Call, D. A. 2007. "An Integrated Hazard Analysis of Ice Storm Impacts, Warnings, and Emergency Plans." PhD diss., Syracuse University.

Call, D. A. 2010. "Changes in Ice Storm Impacts over Time: 1886–2000." *Weather, Climate, and Society* 2 (1): 23–35. https://doi.org/10.1175/2009WCAS1013.1.

Call, D. A., K. E. Grove, and P. J. Kocin. 2015. "A Meteorological and Social Comparison of the New England Blizzards of 1978 and 2013." *Journal of Operational Meteorology* 3 (1): 1–10. https://doi.org/10.15191/nwajom.2015 .0301.

Centers for Disease Control. 2013. "Deaths Associated with Hurricane Sandy— October–November 2012." *Morbidity and Mortality Weekly Reort* 62 (20): 393–97. https://www.cdc.gov/mmwr/preview/mmwrhtml/mm6220a1.htm.

Changnon, D., C. Merinsky, and M. Lawson, 2008. "Climatology of Surface Cyclone Tracks Associated with Large Central and Eastern U.S. Snowstorms, 1950–2000." *Monthly Weather Review* 136, 3193–3202, https://doi.org/10.1175 /2008MWR2324.1.

Changnon, S. A. 2011. "Windstorms in the United States." *Natural Hazards* 59, 1175–87. https://doi.org/10.1007/s11069-011-9828-2.

Charney, J. G. 1947. "The Dynamics of Long Waves in a Baroclinic Westerly Current." *Journal of Meteorology* 4, 135–62. https://doi.org/10.1177/03091 33309339797.

Charney, J. G. 1949. "On a Physical Basis for Numerical Prediction of Large-Scale Motions in the Atmosphere." *Journal of Meteorology* 6, 371–85.

Charney, J. G. 1954. "Numerical Prediction of Cyclogenesis." *Proceedings of the National Academy of Sciences* 40, 99–110. https://doi.org/10.1073/pnas .40.2.99.

Charney, J. G. 1955. "Numerical Methods in Dynamic Meteorology." *Proceedings of the National Academy of Sciences* 41, 798–802.

Charney, J. G., and N. A. Phillips. 1953. "Numerical Integration of the Quasi-Geostrophic Equations for Barotropic and Simple Baroclinic Flows." *Journal of Meteorology* 10 (2): 71–99. https://doi.org/10.1175/1520-0469(1953)010 <0071:NIOTQG>2.0.CO;2.

Cox, J. D. 2002. *Storm Watchers: The Turbulent History of Weather Prediction from Franklin's Kite to El Nino.* Wiley.

Cutter, S. L., B. J. Boruff, and W. L. Shirley. 2003. "Social Vulnerability to Envi-

ronmental Hazards." *Social Science Quarterly* 84 (2): 242–61. https://doi
.org/10.1111/1540-6237.8402002.

Cutter, S. L., ed. 2006. *Hazards, Vulnerability and Environmental Justice.* Routledge.

Dahl, T. E., and S. M. Stedman. 2013. *Status and Trends of Wetlands in the
Coastal Watersheds of the Conterminous United States 2004 to 2009.* U.S.
Department of the Interior, Fish and Wildlife Service, and National Oceanic
and Atmospheric Administration, National Marine Fisheries Service.
https://www.fws.gov/wetlands/documents/status-and-trends-of-wetlands
-in-the-coastal-watersheds-of-the-conterminous-us-2004-to-2009.pdf.

Delaney, B. 2000. "The First Homeowner's Policies Are Offered." In *Landmarks
in Modern American Business*, 341–47. Salem Press.

Dixon, P. G., D. M. Brommer, B. C. Hedquist, A. J. Kalkstein, G. B. Goodrich,
J. C. Walter, C. C. Dickerson IV, S. J. Penny, and R. S. Cerveny. 2005. "Heat
Mortality versus Cold Mortality: A Study of Conflicting Databases in the
United States." *Bulletin of the American Meteorological Society* 86 (7): 937–
44. https://doi.org/10.1175/BAMS-86-7-937.

Endlich, R. M. 1953. "A Study of Vertical Velocities in the Vicinity of Jet Streams."
Journal of Meteorology 10, 407-415.

Federal Emergency Management Agency. 2013. *Hurricane Sandy After-Action
Report.* FEMA.

Fleming, J. R. 2016: *Inventing Atmospheric Science: Bjerknes, Rossby, Wexler,
and the Foundations of Modern Meteorology.* MIT Press.

Gilchrist, B. 2006. "Remembering Some Early Computers, 1948–1960." Personal memoir associated with Columbia University EPIC. https://web
.archive.org/web/20121022185315/http://www.columbia.edu/cu/epic/pdf
/gilchrist_3.07.06.pdf.

Grumm, R. 2010a. *The Historic Storm of 24–26 November 1950.* Penn State SREF
case study series. http://cms.met.psu.edu/sref/severe/1950/26Nov1950.pdf.

Grumm, R. 2010b. *Snowmelt and Rain Floods of 4–9 December 1950.* Penn State
SREF case study series. http://cms.met.psu.edu/sref/severe/1950/04Dec
1950.pdf.

Halverson, J. B., and T. Rabenhorst. 2013. "Hurricane Sandy: The Science and
Impacts of a Superstorm." *Weatherwise* 66 (2): 14–23. https://doi.org/10
.1080/00431672.2013.762838.

Hampson, N. B. 2016. "U.S. Mortality Due to Carbon Monoxide Poisoning, 1999–2014, Accidental and Intentional Deaths." *Annals of the American Thoracic Society* 13, 1768–74.

Haney, C. R. 2017. "Winter Weather Hazards: Injuries and Fatalities Associated with Snow Removal." PhD diss., Mississippi State University. https://scholars junction.msstate.edu/td/5028.

Harris, D. L., and C. V. Lindsay. 1957. *An Index of Tide Gages and Tide Gage Records for the Atlantic and Gulf Coasts of the United States.* National Hurricane Research Project Report No. 7.

Hart, R. E., and R. H. Grumm. 2001. "Using Normalized Climatological Anomalies to Rank Synoptic-Scale Events Objectively." *Monthly Weather Review* 129 (9): 2426–42. https://doi.org/10.1175/1520-0493(2001)129<2426 :UNCATR>2.0.CO;2.

Hunt, Jr., F. J. 1962. "Homeowners—The First Decade." *Proceedings of the Casualty Actuary Society* 49, 12–36.

Insurance Institute for Highway Safety. 2009. "IIHS's 50th Anniversary: Crashworthiness Then and Now." https://www.iihs.org/about-us/50th-anniversary

James, R. W. 1952. "The Latitude Dependence of Intensity in Cyclones and Anticyclones." *Journal of Meteorology* 9, 243–51.

Jenkin, P. 1978. "Suspected of Murder in Marblehead." *Yankee,* January 1978. https://newengland.com/today/living/new-england-history/murder -marblehead/.

Karl, T. R., and W. J. Koss. 1984. *Regional and National Monthly, Seasonal, and Annual Temperature Weighted by Area, 1895–1983.* Historical Climatology Series 4-3. National Climatic Data Center. https://www.regulations.gov /document/EPA-HQ-OAR-2013-0566-0218.

Karl, T. R., J. M. Melillo, and T. C. Peterson (eds.). 2009. *Global Climate Change Impacts in the United States.* Cambridge University Press.

Kelly, J. 2008. "It's Not the Eggnog Talking: Christmas Is Starting Earlier." *Washington Post,* November 20, 2008. http://www.washingtonpost.com wp-dyn/content/article/2008/11/19/AR2008111903883.html.

Kistler, R., E. Kalnay, W. Collins, S. Saha, G. White, J. Woollen, M. Chelliah, et al. 2001. "The NCEP-NCAR 50-Year Reanalysis: Monthly Means CD-ROM and Documentation." *Bulletin of the American Meteorological Society* 82 (2): 247–67.

Kistler, R. E., L. Uccellini, and P. J. Kocin. 2004. "Thanksgiving Weekend Storm of 1950." Paper presented at the Norm Phillips Symposium, Seattle, WA, American Meteorological Society, January 2004. Abstract, sources, and recording available at https://ams.confex.com/ams/84Annual/techprogram/paper_73168.htm.

Kneeland, T. W. 2020. *Playing Politics with Natural Disaster: Hurricane Agnes, the 1972 Election, and the Origins of FEMA.* Cornell University Press.

Kneeland, T. W. 2021. *Declaring Disaster: Buffalo's Blizzard of '77 and the Creation of FEMA.* Syracuse University Press.

Kocin, P. J., P. N. Schumacher, R. F. Morales, and L. W. Uccellini. 1995. "Overview of the 12–14 March 1993 Superstorm." *Bulletin of the American Meteorological Society* 76, 165–82.

Kocin, P. J., and L. W. Uccellini. 2004. *Northeast Snowstorms.* Meteorological Monographs, vol. 32, no. 54. American Meteorological Society.

Lange, F., A. L. Olmstead, and P. W. Rhode. 2009. "The Impact of the Boll Weevil, 1892–1932." *The Journal of Economic History* 69 (3): 685–718. https://doi.org/10.1017/S0022050709001090.

Lindop, E., and S. De Capua. 2009. *America in the 1950s.* Twenty-First Century Books.

Ludlum, D. M. 1982. *The American Weather Book.* Houghton Mifflin Company.

Lynch, P. 2008. "The ENIAC Forecasts: a Re-creation." *Bulletin of the American Meteorological Society* 89 (1): 45–55.

McCartney, S. 2001. *ENIAC: The Triumphs and Tragedies of the World's First Computer.* Reissue edition. Berkley Trade.

Meteorological Development Laboratory. 2019. "The History of the Meteorological Development Laboratory." https://vlab.noaa.gov/web/mdl/history.

Miller, D., and A. Reeves. 2021. "Pass the Buck or the Buck Stops Here? The Public Costs of Claiming and Deflecting Blame in Managing Crises." *Journal of Public Policy* 11, 1–29. https://doi.org/10.1017/S0143814X21000039.

Mook, C. P. 1949. "The Famous Storm of November, 1913." *Weatherwise* 2 (6): 126–28.

Morrill, R. 2015. "Fifty Years of US Poverty: 1960–2010." February 19, 2015. http://www.newgeography.com/content/004852-50-years-us-poverty-1960-2010.

Nagashybayeva, G. 2009. "Homeowners Insurance." Library of Congress. Last updated January 25, 2021. https://www.loc.gov/rr/business/businesshistory/September/homeowners_ins.html.

National Weather Service. 1994. *Superstorm of March 1993, March 12–14, 1993.* Natural Disaster Survey Report. National Oceanic and Atmospheric Administration, U.S. Department of Commerce. https://www.weather.gov/media/publications/assessments/Superstorm_March-93.pdf.

National Weather Service. 1998. *The Ice Storm and Flood of January 1998.* Service assessment. National Oceanic and Atmospheric Administration, U.S. Department of Commerce. https://www.weather.gov/media/publications/assessments/iceflood.pdf.

Perry, C. A. 2000. *Significant Floods in the United States during the 20th Century—USGS Measures a Century of Floods.* USGS Fact Sheet 024-00. U.S. Department of the Interior. https://doi.org/10.3133/fs02400.

Phillips, N. A. 1951. "A Simple Three-Dimensional Model for the Study of Large-Scale Extratropical Flow Patterns." *Journal of Meteorology* 8, 381–94.

Phillips, N. A. 1958. "Geostrophic Errors in Predicting the Appalachian Storm of November 1950." *Geophysica* 6, 389–405.

Phillips, N. A. 2001. "The Start of Numerical Weather Prediction in the United States." In *50th Anniversary of Numerical Weather Prediction Comemorative Symposium*, edited by A. Spekat, 13–28. Deutsche Meteorologische Gesellschaft.

Phillips, N. A. 2004. "NWP and the Appalachian Storm of November 1950." Keynote presentation at the Symposium on the 50th Anniversary of Operational Numerical Weather Prediction, June 2004, University of Maryland.

Pickenpaugh, R. 2001: *Buckeye Blizzard: Ohio and the 1950 Thanksgiving Storm.* Gateway Press.

Pielke, R. A., Jr., J. Gratz, C. W. Landsea, D. Collins, M. Saunders, and R. Musulin. 2008. "Normalized Hurricane Damages in the United States: 1900–

2005." *National Hazards Review* 9, 29–42. https://www.nhc.noaa.gov/pdf /NormalizedHurricane2008.pdf.

Pore, N. A., and C. S. Barrientos. 1976. *Storm Surge.* MESA New York Bight Atlas Monograph 6. New York Sea Grant Institute.

Pore, N. A., W. S. Richardson, and H. P. Perrotti. 1974. *Forecasting Extratropical Storm Surges for the Northeast Coast of the United States.* NOAA Technical Memorandum NWS TDL-50. U.S. Department of Commerce.

Rattigan, D. 2005. "1950 Slaying Case Still Gripping," *Boston Globe,* November 30, 2005. http://archive.boston.com/news/local/articles/200510/30 /1950_slaying_case_still_gripping/.

Richardson, L. F. 1922. *Weather Prediction by Numerical Process.* Cambridge University Press.

Richardson, W. S., and C. L. Boggio. 1980. *A New Extratropical Storm Surge Forecast Equation for Charleston, South Carolina.* NWS TDL Office Note 80-7. U.S. Department of Commerce. https://repository.library.noaa.gov /view/noaa/6780.

Richardson, W. S., and C. S. Gilman. 1983a. *Improved 6-, 12-, 18-, and 24-h Extratropical Storm Surge Forecast Guidance for Boston, Mass.; New York, N.Y.; Norfolk, Va.; and Charleston, S.C.* NWS TDL Office Note 83-8. U.S. Department of Commerce. https://www.nws.noaa.gov/mdl/pubs/Doc uments/OfficeNotes/MDL_OfficeNote83-8.pdf.

Richardson, W. S., and C. S. Gilman. 1983b. *Improved 6-, 12-, 18-, and 24-h Extratropical Storm Surge Forecast Guidance for Willets Point, N.Y.* NWS TDL Office Note 83-17. U.S. Department of Commerce. https://www.nws .noaa.gov/mdl/pubs/Documents/OfficeNotes/MDL_OfficeNote83-17.pdf.

Sanders, F., and J. R. Gyakum. 1980. "Synoptic–Dynamic Climatology of the 'Bomb.'" *Monthly Weather Review* 108 (10): 1589–1606. https://doi.org/10 .1175/1520-0493(1980)108<1589:SDCOT>2.0.CO;2.

Schmidlin, T. W., and J. A. Schmidlin. 1996. *Thunder in the Heartland: A Chronicle of Outstanding Weather Events in Ohio.* Kent State University Press.

Schwartz, R. 2007. *Hurricanes and the Middle Atlantic States.* Blue Diamond Books.

Shoup, D. 2005. *The High Cost of Free Parking*. Planners Press, American Planning Association.

Smith, C. D., Jr. 1950. "The Destructive Storm of November 25–27, 1950." *Monthly Weather Review* 78, 204–9.

Smith, C. D., Jr., and C. L. Roe. 1952. "Comparisons between the Storms of November 20–22, 1952, and November 25–27, 1950. *Monthly Weather Review* 80, 227–31.

Squires, M. F., J. H. Lawrimore, R. R. Heim Jr., D. A. Robinson, M. R. Gerbush, and T. W. Estilow. 2014. "The Regional Snowfall Index." *Bulletin of the American Meteorological Society* 95, 1835–48. https://doi.org/10.1175/BAMS-D-13-00101.1.

Sylves, R. 2020. *Disaster Policy and Politics: Emergency Management and Homeland Security*. CQ Press.

Tolle, M. E. 2012. *What Killed Downtown? Norristown, Pennsylvania, from Main Street to the Malls*. Self-published.

Uccellini, L. W., P. J. Kocin, R. S. Schneider, P. M. Stokols, and R. A. Dorr. 1995. "Forecasting the 12–14 March 1993 Superstorm." *Bulletin of the American Meteorological Society* 76, 183–200.

U.S. Bureau of the Census. 1953. *Census of Housing: 1950*. https://www.census.gov/library/publications/1953/dec/housing-vol-01.html.

U.S. Energy Information Administration. 2018. *What's New in How We Use Energy at Home: Results from EIA's 2015 Residential Energy Consumption Survey*. U.S. Department of Energy. https://www.eia.gov/consumption/residential/reports/2015/overview/pdf/whatsnew_home_energy_use.pdf.

Wagenmaker, R., and G. Mann. 2013. *The "White Hurricane" Storm of November 1913: A Numerical Model Retrospective*. https://www.crh.noaa.gov/images/dtx/climate/1913Retrospective.pdf.

Walz, F. J. 1913. "Climatological Data for November, 1913: District 3, the Ohio Valley." *Monthly Weather Review* 41, 1665–77.

Works Progress Administration. 1939. *The WPA Guide to 1930s New Jersey*. Viking Press.

Yarbrough, T. E. 1987. *A Passion for Justice: J. Waties Waring and Civil Rights*. Oxford University Press.

Zeller, K. 1998. "Mountainlair Turns 50." *West Virginia University Alumni Magazine* 21 (2).

Newspaper sources (names in November 1950)

Altoona Mirror

Altoona Tribune

Asbury Park Press

Asheville Citizen

Atlanta Journal

Baltimore Sun

Bangor News

Berkshire Eagle (Pittsfield, Mass.)

Birmingham News

Bloomington (Ill.) Pantagraph

Bluefield (W.Va.) Daily Telegraph

Boston Globe

Bradford Era

Bristol (Va.) Herald Courier

Buffalo Evening News

Burlington Free Press

Butler (Pa.) Eagle

Charleston (W.Va.) Gazette

Charleston (S.C.) News and Courier

Charlotte Observer

Chattanooga Times

Chicago Tribune

Cincinnati Enquirer

Clarksburg Telegram

Clearfield Progress

Cleveland Plain Dealer

Columbus (Ohio) Dispatch

Dothan Eagle

Erie Times

Evansville Courier

Florence News

Florida Times Union (Jacksonville)

Fort Wayne News-Sentinel

Greenville (Miss.) Delta-Democrat Times

Greenville (S.C.) News

Harrisburg Evening News

Harrisburg Patriot

Harrisonburg Daily News-Record

Hartford Courant

Hattiesburg American

Huntsville Times

Indianapolis Star

Jackson (Miss.) Clarion Ledger

Knoxville News Sentinel

Lexington Herald

Louisville Courier Journal

Manchester (N.H.) Morning Union

Mansfield (Ohio) News Journal

Marietta (Ohio) Times

Memphis Commercial Appeal

Miami Herald

Montgomery Advertiser

Nashville Tennessean

New York Times

Oil City Blizzard

Orlando Evening Star

Orlando Sentinel

Ottawa (Ontario) Journal

Philadelphia Inquirer

Pittsburgh Press

Portland (Maine) Press Herald/Sunday Telegram

Providence Journal

Raleigh News and Observer

Richmond Times Dispatch

Roanoke Times

Savannah News

Scranton Tribune/Scrantonian

Southern Illinoisian (Carbondale)

The State (Columbia, S.C.)

Steubenville Herald Star

Syracuse Herald-Journal

Tallahassee Democrat

Toledo Blade

Troy Times-Record

Tupelo Journal

Vineland Times-Journal

Virginian Pilot (Norfolk)

Wheeling News-Register

Wilmington (Del.) Journal-Every Evening

Winston-Salem Journal

Youngstown Vindicator

Additional newspapers were consulted for more details, context, or cross-referencing of facts, but they were not systematically examined.

Weather data sources

Climatological reports by the Weather Bureau, U.S. Department of Commerce (including narrative reports of impacts):

    Climatological Data, National Summary, Annual 1950

    Climatological Data, National Summary, November 1950

    Climatological Data, [individual states/regions], November 1950

    Weekly Weather and Crop Bulletin for the weeks ending November 28, December 5, and December 12, 1950

Daily weather maps produced by the Weather Bureau, U.S. Department of Commerce

Newspaper accounts

Weather observations from archives maintained by the National Centers for Environmental Information, National Oceanic and Atmospheric Administration, U.S. Department of Commerce

INDEX

ABOUT THE AUTHOR

DAVID A. CALL IS AN ASSOCIATE PROFESSOR OF GEOGRAPHY AND METEO-
rology at Ball State University. He received his meteorology degree
with honors from Penn State and advanced degrees in geography from
Syracuse University. Call teaches classes in meteorology and physical geography, including a storm chasing class where he takes students to the Great Plains in search of tornadoes. His research examines the impacts of hazardous winter weather and has been published
in the *Professional Geographer, Bulletin of the American Meteorological
Society*, and elsewhere. He lives in Muncie, Indiana, with his family,
who graciously tolerate his traffic light collection.